Cosmology and Creation

Cosmology and Creation

The Spiritual Significance of Contemporary Cosmology

Paul Brockelman

New York   Oxford
Oxford University Press
1999

Oxford University Press

Oxford   New York
Athens   Auckland   Bangkok   Bogotá   Buenos Aires   Calcutta
Cape Town   Chennai   Dar es Salaam   Delhi   Florence   Hong Kong   Istanbul
Karachi   Kuala Lumpur   Madrid   Melbourne   Mexico City   Mumbai
Nairobi   Paris   São Paulo   Singapore   Taipei   Tokyo   Toronto   Warsaw

and associated companies in
Berlin   Ibadan

Published by Oxford University Press, Inc.
198 Madison Avenue, New York, New York 10016

Oxford is a registered trademark of Oxford University Press, Inc.

Library of Congress Cataloging-in-Publication Data
Brockelman, Paul T.
Cosmology and creation: the spiritual significance of
contemporary cosmology/Paul Brockelman.
p. cm.
Includes bibliographical references and index.
ISBN 0-19-511990-8 (cloth)
1. Creation. 2. Cosmology. I. Title.
BL226.B76 1999
215'.2--dc21   98-28273

Design by Adam B. Bohannon
Set in Joanna and Gill

9 8 7 6 5 4 3 2 1
Printed in the United States of America
on acid-free paper

For
Barbara, Tom, and Mira
and my brother, Henry.

Contents

Preface

It seems that the great age of geographical exploration has ended. We have seen it all: We have mapped the continents, sailed all the seas, climbed the highest mountains, and walked the dry deserts. Of course the floor of the sea remains mostly unexplored as do the heavens above, especially beyond our home within the Solar System. In any case, although there is still much to learn about our old earth, the intellectual cutting edge seems to have shifted to cosmology—not just the beginnings, but the whole immense story which science has provided us in the past few decades concerning the unfolding of the universe over some fifteen billion years. It is an astonishing story of scientific exploration and discovery with—I believe—great cultural, moral, and religious significance for our time.

Indeed, it is in terms of the scientific exploration of the heavens that we have recently witnessed tremendous growth in our understanding, an exploratory understanding which promises to continue at the same pace in the near future. With the development of radio and X-ray astronomy, spectroscopy, the amazing technology of very fine telescopes like the new Hubble Space Telescope, as well as cyclotrons and particle physics to get at the most minute bits of matter, we have pushed back in time and out into space our comprehension of the universe. The science which constitutes this deepened understanding is cosmology: the study of the origins and development of the universe over twelve to fifteen billion years.

But the intellectual exploration is not just in physics, astronomy, biology, and the other sciences. There is also a cross-disciplinary, intellectual exploration and conversation taking place which has to do with the philosophical, moral, and religious significance of these breakthroughs. And this has to do with their bearing on how we see reality and life today and how we see our role and destiny within it. Indeed, some have suggested that the new cosmology as well as the new kinds of science (e.g., quantum physics and chaos theory) that have advanced it promise to change the everyday symbolic universe or intellectual framework of understanding in which we find ourselves. In other words, we may be emerging into a new cultural and intellectual era, an era which will see life and our place within it in as radically different a way as the Enlightenment worldview was from the Medieval view which it replaced.

Imagine my excitement then when in 1995 my astrophysicist colleague and friend Eberhard Moebius (of the Earth, Ocean, and Space Center at the University of New Hampshire) invited me to coteach with him a seminar in the Physics Department to explore the implications of this modern cosmology and the sciences which have led to it. The seminar was composed of about fifteen students—undergraduates in physics, philosophy, and biology; and graduate students and young instructors from some of the sciences who were interested in the broader intellectual questions which the cosmology has suggested.

While growing up in Germany, Eberhard had felt the clash between contemporary science and the sort of creation myths involved in Genesis, a clash so typical of modern culture over the past few centuries. The experience left him hungry for a way to heal the breach between science and religion, between his head and his heart.

I, on the other hand, had grown up in a New England Unitarian family and had come to teach philosophy and religious studies at UNH. Particularly because of my interest in contributing to the amelioration of environmental difficulties, I felt that we needed to renew our spiritual sense of reverence for nature as *intrinsically* valuable in itself and not merely *extrinsically* valuable in so far as it provides "resources" which in a utilitarian manner we can turn into "useful" products for the consumer industrial societies.

The seminar was hugely successful because all of us were interested in these bigger questions and found too little opportunity in the context of traditional departmental approaches and offerings to deal with them. As a reporter who sat in on the seminar put it,

> during the course of the semester, as they discussed topics such as "What is Reality?" "Science and Mysticism" and "The Theory of Everything," it became clear that the philosopher and the physicist had much in common. By the end, they were sometimes finishing each other's sentences.[1]

Our exploration led us to a very interesting conversation concerning the relationship between contemporary science and religion. It seemed to us that not only were the two compatible, but that they in fact seemed to mutually support one another in so far as they lead ultimately to mysticism. That doesn't mean ignorance or mumbo jumbo, nor does it mean that we ought to give up the scientific endeavor to understand. Rather, mysticism is a form of spiritual life which many twentieth-century scientists such as Albert Einstein and Irwin Schroedinger endorsed, and which many scholars believe lies at the core of human religious life in general.

This seminar, then, was the immediate background and context for the book which follows. Much of the perspective and many of the arguments it contains were suggested or shaped by the probing explorations and conversations which developed within it. I am deeply indebted to my colleague Eberhard Moebius and the participants in that seminar for their help; but of course I am personally responsible for the views espoused within the following essay and for whatever errors it may contain.

Cosmology and Creation

We do not know whether we shall succeed in once more expressing the spiritual form of our future communities in the old religious language. A rationalistic play with words and concepts is of little assistance here; the most important preconditions are honesty and directness. But since ethics is the basis for the communal life of men, and ethics can only be derived from that fundamental human attitude which I have called the spiritual pattern of the community, we must bend all efforts to reuniting ourselves, along with the younger generation, in a common human outlook. I am convinced that we can succeed in this if again we can find the right balance between the two kinds of truth.

§ Werner Heisenberg, Physics and Beyond

Contrary to the strict division of the activity of the human spirit into separate departments—a division prevailing since the nineteenth century—I consider the ambition of overcoming opposites, including also a synthesis embracing both rational understanding and the mystical experience of unity, to be the mythos, spoken and unspoken, of our present day and age.

§ Wolfgang Pauli, Writings on Physics and Philosophy

In both science and religion, we seek creation myths, stories that give our lives meaning.

§ George Johnson, Fire in the Mind

Introduction
The Unfurnished Eye

Are We Losing Touch with Reality?

I remember those warm summer evenings of my youth when, finished with after-supper playing and before going home for the night, my friends and I would lie on our backs on the grass and gaze up in wonder at the infinity appearing before our eyes as the stars and galaxies emerged in the unfolding night. As I recall, the evening breeze had diminished to a near hush, the air itself was still warm, soft, and full of the moist perfumes of pine and freshly mowed grass.

We would quiet ourselves down as we pointed out emerging star figures—there is the big dipper, there are the Pleiades and the evening star—just as had countless human generations before us. The ancient world, we are told by scholars of those past times and places, thought they were seeing heaven and the gods themselves, or at least the lights of heaven shining through the canopy of the sky to impress us with its mysterious and wonderful sacredness. We were full of wonder, for this was "seeing" things in a new and different way, seeing nature and life itself not as merely a local and practical affair (although of course it is that too), but as a kind of epiphany in which we felt part of a much larger and more interesting reality than we were conscious of while playing earlier. That earlier reality was there to run on, to hide behind, to use to hit the ball. But the reality we encountered there on our backs as the night came on, the reality we perceived through the billions of stars hurled infinitely across the black sky, was less useful, more mysterious, an encompassing universe of which we were part that took our breaths away. We felt wonder and amazement before it, and gratitude too that we were there ourselves and aware of the remarkable reality to which we belonged. It was a different and enthralling way of seeing things.

Even a materialist such as Richard Leakey has observed that our human nature is finally defined by this sense of awe in the face of the night sky:

One does not have to be especially spiritual to experience awe at the infinity of galaxies we can see in the night sky. Our human consciousness does not merely make possible the question Why? It insists that the question be asked. The urge to know is a defining feature of humanity: to know about the past; to understand the present; to glimpse what the future may hold. As Arnold Toynbee said of the impact of subjective consciousness on Homo Sapiens, "This spiritual endowment of his condemns him to a lifelong struggle to reconcile himself with the universe into which he has been born." The night sky is full of unanswered questions.[1]

Whether American or Russian, the astronauts who first ventured into space and ultimately travelled to the moon and back experienced something just like that when they looked back at earth which, as James Irwin put it, "shrank to the size of a marble, the most beautiful marble you can imagine."[2] The Russian cosmonaut, Olag Makarov (hardly a conventionally religious person), insisted that the sight of earth from space made everyone feel a sense of wonder:

> It didn't matter whether the cosmonaut was on a one-man mission in the first Vostoks or part of a large crew on a mission in a modern Soyuz, no one has been able to restrain his heartfelt wonder at the sight of the enthralling panorama of the Earth.[3]

Seeing the earth this way was a spiritual experience in which the planet was encountered as divine, or at least as a remarkable manifestation of the divine. It was no longer just a smooth functioning machine, a backdrop to the human drama, a pile of natural resources put here for our self-centered use, a mere "thing" or object with utility for our endless satisfaction and exploitation. Now the earth was seen as a pulsing and fragile home to which we belong, a mighty, beautiful and fragile creation fresh from the hand of God. As the astronaut Edgar Mitchell says:

> Instead of an intellectual search, there was suddenly a very deep gut feeling that something was different. It

occurred when looking at earth and seeing this blue-and-white planet floating there, and knowing it was orbiting the sun, seeing that sun, seeing it set in the background of the very deep black and velvety cosmos, seeing—rather knowing for sure—that there was a purposefulness of flow, of energy, of time, of space in the cosmos—that it was beyond man's rational ability to understand, that suddenly there was a nonrational way of understanding that had been beyond my previous experience.

There seems to be more to the universe than random, chaotic, purposeless movement of a collection of molecular particles.

On the return trip home, gazing through 240,000 miles of space toward the stars and the planet from which I had come, I suddenly experienced the universe as intelligent, loving, harmonious.... My view of our planet was a glimpse of divinity.[4]

The wonder these astronauts felt afforded a new way of seeing things, a way of seeing that many think may lead to an entirely new vision of encompassing nature and our place within it. In his introduction to *The Home Planet*, the editor Kevin Kelley puts it this way:

Space offers us a chance to see our world with new eyes, a perspective that may have great significance for the planet for all of the future.... I think this sense of wonder at our universe and the strangeness of our lives within our tiny part of it is important to our sense of ourselves and perhaps to our very survival.[5]

But how rarely do we perceive the universe and life itself that way! In our ordinary lives most of us experience being as a sort of "unextraordinary" (that is to say ordinary) and practical reality in which we carry out our everyday activities. It's almost as if our normal condition is a benumbed and altered state of consciousness, lost in a world in which we perceive things as humdrum and unremarkable, unconscious that the very fact that they exist—that "anything" is at all and that there are so many different forms of it—is truly extraordinary. Add to this numbness the hectic pace and separation if not isolation from nature which our consumer industrial societies foster and it seems possible that we are in danger of losing touch with any larger and more meaningful reality than ourselves. If "in vitro" in science means an experiment "under glass" in the laboratory rather than in nature, then John Fowles argues that "the evolution of human mentality has put us all in vitro now, behind the glass wall of our own ingenuity."[6]

In the nineteenth century Emerson described such a deadening of our ability to see things in their spiritual depth in his famous essay, "Nature."

> [F]ew adults can see nature. Most persons do not see the sun, at least they have a very superficial seeing. The sun illuminates only the eye of the man, but shines into the eye and the heart of the child.[7]

It's as if our vision becomes clouded and hazy as we grow up, till by the time we have become adults seriously responsible for ourselves and others it barely exists and any sense of wonder at the epiphany of nature is nearly or at least normally absent. William Blake tells us that "if the doors of perception were cleansed everything would appear as it is, infinite."[8]

The Felt Need for Something More

A number of contemporary thinkers have argued that this benumbing of our spiritual ability to perceive the infinite within nature is a kind of cultural illness or myopia which is one of the central issues of our time. Vaclav Havel, for example, has pointed to a modern condition which he believes is spiritually painful and empty, socially destablizing, and environmentally disastrous. Our modern cultures, which evolved out of the European Enlightenment and which by now have spread around the world have led to the demise of the traditional religious sense of a divine creator in favor of an anthropocentric and myopic focus on ourselves and our immediate practical needs and satisfactions. Such a state, Havel believes, is profoundly demoralizing.

> We live in an age in which there is a general turning away from Being: our civilization, founded on a grand upsurge of science and technology, those great intellectual guides on how to conquer the world at the cost of losing touch with Being, transforms man its proud creator into a slave of his consumer needs.... A person who has been seduced by the consumer value system, whose identity is dissolved in an amalgam of the accoutrements of mass civilization, and who has no roots in the wider order of Being, no sense of responsibility for any higher reality than his or her own personal survival, is a demoralized person and, by extension, a demoralized society.[9]

For Havel, then, the absence of any sense of the transcendental value of the universe—the loss of any connection to a "wider order of Being"—constitutes what he terms a demoralized life. It is demoralized spiritually in that instead of a sense of a wider reality of which we are part and in which we see our role and destiny in life we limit ourselves merely to fulfilling our own often pathetically narrow and superficial interests and desires. And this of course entails an ethical demoralization because what is of supreme value in such a state is simply relative to human needs and longing. Who we are and what we want to get out of life gets reduced to passing fads and fancies, lacking any larger vision of the whole.

In a recent Fourth of July speech at Independence Hall in Philadelphia at which he was awarded the Liberty Medal, Havel went on to indicate that this demoralized culture is the result of a misperception of reality which leads to a profoundly alienated state of being.

[T]he relationship to the world that modern science fostered and shaped appears to have exhausted its potential. The relationship is missing something. It fails to connect with the most intrinsic nature of reality and with natural human experience. It produces a state of schizophrenia: man as an observer is becoming completely alienated from himself as a being.[10]

We become disoriented, then, left to survive in a world without a spiritual compass and stripped of any framework of values. But people need spiritual meaning and purpose in order to feel fulfilled. They need to be connected to a wider encompassing reality in which they can get their bearings.[11]

In this sense, our modern industrial world seems surprisingly like the Roman world of Palestine at the time of Jesus. New Testament scholar Burton Mack has characterized that world as one in which earlier nations and cultural traditions were being absorbed into the larger political economy of the Hellenistic and Roman period. That meant not only an intermingling of those earlier religious views and practices through trade and the massive transplantation of peoples from one place to another, but also the shattering of those earlier traditions as simply too parochial, too limited to include all of the world that was coming to be. The breakdown of those earlier and more parochial spiritual visions of life brought with it a great sense of spiritual disorientation, if not utter meaninglessness and despair. New forms of philosophical and religious vision and practice arose to meet this situation, and ultimately, of course, it was the Christ cult that finally won the day by providing a more encompassing view of the world and the place of the various cultural, ethnic, and religious traditions within it.[12]

This, of course, is not an argument for Christianity, but our contemporary culture faces some of the same conditions and issues as the Roman empire—a breakdown of nation states, the emergence of a world economic order, and the fracturing of local religious traditions and practices.

It would seem that we are witnessing something quite like

that in our own time. Political thinker David Bollier has argued that our culture must come to grips with just such pervasive ethical and spiritual emptiness.

> The truth is, Americans in the late twentieth century need more than the First Amendment and its case law to bind them together. They need a new cultural covenant with each other that can begin frankly to address the spiritual void in modern secular society.[13]

In other words, Bollier tells us that people want and need both a sense of meaning and a sense of community in life.

Havel thinks that the lack of awareness and appreciation for any "wider order of Being," as he puts it, challenges our postmodern world to develop a new spiritual relationship to the broader reality of the universe from which we have been generated and in which we are sustained.[14]

In his 1995 Commencement address at Harvard, Havel neatly summarized his views.

> The main task in the coming era is...a radical renewal of our sense of responsibility. Our conscience must catch up to our reason, otherwise we are lost.
>
> It is my profound belief that there is only one way to achieve this: we must divest ourselves of our egoistic anthropocentrism, our habit of seeing ourselves as masters of the universe who can do whatever occurs to us. We must discover a new respect for what transcends us: for the universe, for the earth, for nature, for life, and for reality. Our respect for other people, for other nations, and for other cultures, can only grow from a humble respect for the cosmic order and from an awareness that we are a part of it, that we share in it and that nothing of what we do is lost, but rather becomes part of the eternal memory of Being, where it is judged....
>
> Whether our world is to be saved from everything that threatens it today depends above all on

whether human beings come to their senses, whether they understand the degree of their responsibility and discover a new relationship to the very miracle of Being.[15]

What we have lost, then, is the ability to see our lives as part of a wider order and reality beyond our daily and passing self-centered desires and dreams. By seeing nature and the entire universe as a "stuff" put here for our endless productive transformation and use, we have reduced reality to a mere extrinsic value for us; it is no longer encountered as intrinsically valuable in itself. As a consequence, we have lost any sense of belonging to a larger and more significant drama and reality. If, as I hope to show in this book, such a wider reality is in fact sacred, then God has died, not so much as an idea than as a daily milieu in which he is simply absent, unseen and unheard.

Furthermore, as many ecologists have pointed out, by seeing nature simply as a backdrop and "stuff" put here merely for our pleasure and endless economic exploitation and growth, we have brought upon ourselves and all of creation a vastly destructive ecological crisis in which we ultimately threaten not only our own lives but those of myriad species around us. It is as if we blinded ourselves so that we could never more perceive truly the sort of immense reality to which we belong, that reality which we as children looking up at the stars or the astronauts far out in space saw so very clearly. Of course, it is our modern industrialized society that has become blind to this wider, wondrous reality, and it is that same society that cries out for the sustenance that such a reality may be able to provide. For the sake of the environment as well as our spiritual lives, we need to change. We need to restore our spiritual vision. We need to renew the age-old experience of the sacred dimension of life; what is at stake is our understanding of the very meaning of existence itself.

This seems all the more true when you consider that we are living in a time of declining income and standard of living for most Americans. At such a time, one would think, questions of quality of life as opposed to quantity of material acquisitions would and should become paramount. In fact, many of our young people are rejecting the dreams of the consumer society in favor of finding alternative spiritual perspectives and values.

With New Eyes

Economist Herman Daly and theologian John Cobb have argued just this in their book, *For the Common Good.*

> [A] sustained willingness to change depends on a love of the earth that human beings once felt strongly, but that has been thinned and demeaned as the land was commodified.... there is a religious depth in myriads of people that can find expression in lives lived appropriately to reality. That depth must be touched and tapped.... If that is done, there is hope.... Our point is that the changes that are now needed in society are at a level that stirs religious passions. The debate will be a religious one whether that is made explicit or not.[16]

What is called for, then, is a new way of seeing things that might help us to live more appropriately within nature, to see things with, in Emily Dickinson's haunting phrase, "an unfurnished eye." We need to feel wonder at the extraordinary miracle of life, at the astonishing epiphany it manifests. We need to be touched and changed to our core.

It may be that our contemporary industrial culture is undergoing just such a transformation in how it sees things, that a paradigm shift is permitting us to see nature and life with new eyes. In fact, for a variety of reasons, I believe that is true.

Each of us has had the experience of coming to see some aspect or event in our lives in a radically different way than before, and every culture to one degree or another certainly undergoes changes in the way they perceive things in their world. But relatively few individuals have gone through the kind of shift in their ultimate framework of significance—the way they see life as a whole—that the astronaut Edgar Mitchell experienced; and the same is true for human cultures. Certainly, the startling and explosive transformation of traditional European Christian culture into what we now call "modernity" or "the modern industrial consumer society" in the seventeenth, eighteenth, and nineteenth centuries is a classic and still (to say the least) vastly influential instance of such a cultural change in our way of seeing things. However unusual and even surprising it may seem, I shall argue here that we are currently undergoing just such a radical and remarkable shift in our cultural worldview, a shift from the assumptions and ways of seeing things which characterize the modern industrial culture to what some have called a postmodern or ecozoic point of view. On the one hand, I will argue that changes in contemporary science and religion are permitting (if not at least in part causing) a paradigm shift in the worldview that pervades the modern industrial cultures. And conversely, changes in our worldview brought about by the spiritual, moral, and ecological stresses of that modern industrial world are certainly contributing to the postmodern interpretation of science and religion that is presently unfolding.

I believe that such a change in how we see nature—a "cleansing of the doors of perception" if you will—is today not only possible, but perhaps even inevitable. This is due to two radical changes in how we understand ourselves and our place in the universe. First of all, the scholarly and philosophical understanding of the human condition—in particular the spiritual dimension—has changed profoundly in the past thirty years. As we shall see this shift has important consequences in particular for our appreciation of creation stories and their focal place within human cultures. Secondly, in a parallel development in the past forty or fifty years, our scientific understanding of the nature of this universe we inhabit has been transformed. In fact, many scientists argue that modern cosmology and the quantum physics on which it rests have led to a paradigm shift of immense proportions. In their view it is a shift in how we "see" nature and our place within it. In other words, it constitutes a reconstruction of the symbolic universe in which we find ourselves. If that is so, then we are indeed entering a rather new world in which science and religion—far from being the conventional opponents if not enemies of one another that they have been for the past three hundred years—may now be able to work together to give birth to a new spiritual outlook and comprehensive guide to living. For too long science and religion have lived in separate and often antagonistic worlds. Science, it was thought, leads us to understanding and truth, but seemingly at the cost of any sense of meaning or worship, a way of living blind to what Havel calls "the miracle of being." Religion (or religions), on the other hand, certainly have provided a sense of purpose and reverence in life, but all too often at the cost of understanding and truth, existing out-of-joint with science. We need to bring together our heads and our hearts by linking the scientific creation story that has so recently emerged to our very deep human need for meaning and spiritual fulfillment in life. We need to bring ourselves as observers and scientific knowers together with ourselves as spiritual and moral agents who directly live and experience life. We need to become whole again.

It's not just that we seem to understand something radically new in the big bang cosmology which has erupted into our contemporary consciousness; it's also that the whole fifteen-billion-year development of the entire cosmos (including, of course, the present world) is a story. From the various scientific fields and perspectives has emerged a single narrative understanding of all of creation and our place within it, a story with deep similarities and parallels to the creation stories of traditional cultures.

In a nutshell, my thesis is that the new scientific cosmology[17] which has emerged over the past fifty years has broad and profound implications for our present situation and possibilities, particularly in the spiritual, moral, and cultural dimensions of our lives. I believe that our contemporary understanding of religious life as well as the wide-angle perspective on all of nature that recent science has given us affords an entirely new way to "see" things. Perhaps this way of seeing can help us find that wider reality or metaphysical order which seems so necessary to overcome the spiritual disorientation and ethical demoralization so characteristic of our culture.

Our post-Cold-War and postmodern world needs to feel part of a wider and more meaningful reality. The obvious question emerges: *What might such a reality be?* In what follows, I will explore the nature of the ultimate and inclusive order of Being which seems to be manifest in the remarkable story of creation we now call the "new cosmology," for I believe that it corresponds rather well with the metaphysical order which we need.

A number of contemporary observers such as Thomas Berry, Jay McDaniel, Paul Davies, and Sallie McFague believe that this new scientific understanding of the universe as a whole has unparalleled and revolutionary implications for our world.

Science writer Paul Davies, for example, argues that although science cannot and does not resolve all religious issues it does have deep implications and significance for our thinking about them. "The new physics has overturned so many commonsense notions of space, time, and matter that no serious religious thinker can ignore it."[18] In any case, it has certainly led for the first time in over three hundred years to a real dialogue between scientists and theologians. That is surely unparalleled and revolutionary. But on top of that, it forces us to rethink the classical, supernatural theistic approach to the concept of God (which envisages God as outside nature) and suggests an alternative conception. As the famous scholar of mythology, Joseph Campbell, has said,

> the old notion of a once-upon-a-time First Cause has given way to something more like an immanent ground of being, transcendent of conceptualization, which is in a continuous act of creation now.[19]

It just may be that such an alternative conception of God can provide the sort of wider reality which our poor world needs so desperately. The question of course is whether such a view of God is consistent with what science is telling us about creation and whether it will stir real feelings of reverence and worship in ordinary people.

Looking Ahead

In order to set the scene for our exploration of the divine aspects that the new cosmology makes available, we will first explore in chapter 1 what contemporary scholars tell us about creation mythology. Far from being mere picturesque stories for children, those creation myths found human cultures and religious practices and tie particular cultures to the sacred which they narratively make available. Surprising enough the new scientific cosmology is also a spiritual creation story which, like those early creation myths, makes available the sort of wider reality Havel is suggesting. If that is the case, of course, it seems to indicate that we are entering a remarkable tectonic cultural shift in how we see life. In chapter 2, we shall briefly outline the story which constitutes the new cosmology and some of the shifts in perspective which it seems to entail. In chapter 3, we shall try to draw out what sort of wider and ultimate sacred reality is alive in the universe and made available through that cosmology by phenomenologically exploring the religious experience of wonder and awe. In chapter 4, we shall explore some of the traditional theological ways of thinking about God and nature that push God out of nature and beyond experience, thereby blinding us to that awesome and wonderful power-to-be that both nature and ourselves manifest. This will lead us to explore the question of the "existence" of God in chapter 5. Finally, in chapters 6, 7, and 8 we shall outline a few of the implications and possibilities for spiritual and ethical transformation which this rather radical shift in cultural perspective and way of seeing things seems to entail, including the possibility of healing the breach between science and religion, the breach between our heads and our hearts.

What we humans are looking for in a creation story is
a way of experiencing the world that will open to us the
transcendent, that informs us and at the same time forms
ourselves within it. That is what people want. This is what
the soul asks for.

§ Joseph Campbell, The Mythological Dimension

Any journalist worth his or her salt knows the real story
today is to define what it means to be spiritual. This is
the biggest story—not only of the decade but of the
century.

§ Bill Moyers, The Power of Myth

This Side of Paradise
Creation Mythology

Religion and Ultimate Reality

It seems that all too often we are spiritually asleep, alienated from what is ultimately real in life because we are so busy with our daily tasks that we cannot "see" it. This divorce from what Havel calls "a wider order of being" leaves us in a state of demoralization and distress, longing for something more, spiritually hungry for a reconnection with that ultimate reality. In actuality, when we do become aware of it, we find that we are "in" it as particular forms and modes of it. That ultimate reality is one, and everything that exists is an aspect or form of it since each and every entity precisely "is." It is the business of our various religious traditions to awaken us to that ultimate reality. As cultural historian Thomas Berry and physicist Brian Swimme put it, communion with the mysterious forces that animate the universe "through the story of the universe [a culture's fundamental creation myth] and ceremonial interaction with the various natural phenomena was the traditional way of activating the larger dimensions of our own human mode of being."[1] It is in fact interesting that as far as we know virtually every human culture that has ever existed has developed creation stories which explain how all of reality has come to be just as it happens to be. These stories provide those cultures with a sense of a larger whole to which they belong.

All religions are aimed at transforming how we see our lives and thus how we live them out. Most of us seem to view reality in terms of our immediate wishes and desires. It's as if each of us thought he or she was what is fundamentally meaningful and ultimately real about living. Our moral stance and behavior, of course, reflect this attitude insofar as they seem rooted in egocentrism if not outright narcissism. As Buddhists would say, we become attached to ourselves as if, somehow, a particular "me" is infinite, eternal, and what ultimately counts in life.

All spiritual traditions involve disciplines and tactics to awaken us to a wider reality beyond ourselves. Often that wider reality is the community of others, a love and caring and compassion for fellow human beings, whether on the personal, community, national or even species level. This way of seeing life, although certainly wider and more meaningful than egocentrism, is not yet ultimately real for those traditions.

For many people, most of the time, that's as far as it goes. But once again the great world religions call for more. The more ultimate and encompassing reality is not this or that person, this or that tribe or people, this or that nation, or even humanity as a whole. Rather, it is nature in its entirety—or rather the fact that nature actually is—that is the ultimate and profound reality from which we emerge and in which we live out our allotted times.

Now, the history of how humanity has envisaged nature is an interesting one. Indigenous peoples tended to see it as kindred spirit, related to and involved with the human tribe or community, but obviously more than just that community.

Classical Persian and Greek Gnosticism saw nature in just the opposite way. Nature was not ultimately real, but a fallen state of materiality and flesh that separates us from an ultimate heavenly reality. Life, for them, became a monstrous nightmare, a vale of tears to be escaped from as quickly as possible. In short, nature was seen as a prison which holds us back from our true destiny beyond.

At other times and in other traditions, human beings became so involved in scratching a living from nature that it often seemed merely a backdrop to their practical efforts to survive. That appears to be true for classical Judaism and Christianity. Aside from a few biblical references to the glory of God as manifested in nature and to God's command to act as responsible stewards over it, nature seems to have been either a threatening force or all but irrelevant to human history and redemption. This attitude fed right into the European Enlightenment and consequent industrial revolution, which seems to conceive of nature as simply a conglomeration of natural "resources" put here for our practical utilitarian use.[2]

Nature, then, was merely of extrinsic value, a lot of things to make our lives easier and to facilitate our salvation.

More recently, of course, there has been a renewed interest in nature conceived of as the "environment." Nature in this view is simply the other—the rest of reality that is beyond ourselves. Implicit within this view is that human culture is something different, nonnatural, something outside and beyond it—something special completely outside the original and real nature. This leads to a rather romantic attitude toward nature which wants it to be conserved in its primitive state from the depredations of human culture.

This is a rather narrow and perhaps even dangerous conception of nature. It still seems to make human ends the meaningful point of life, and it thinks of nature as simply the esthetic means to achieve those ends.

My point is that the great spiritual traditions urge us to see a more ultimate and meaningful reality beyond the individual, the culture to which he or she belongs, or even nature pictured as a utilitarian backdrop to the human endeavor. Those traditions express this in different ways, but in effect they are all saying that ultimate reality is something to which both the natural environment and human cultures belong. The renowned philosopher of religion John Hick, for example, argues that the various world religious traditions share a sense of

> transcendence of the ego point of view and its replacement by devotion to or centered concentration upon some manifestation of the Real, response to which produces compassion/love towards other human beings or towards all life.[3]

Whether expressed as Brahman, the Tao, the Dharmakaya, or the Trinity, this ultimate reality to which we are called is neither a thing (nor all of those things) nor a person, but a transcendent and encompassing power-to-be in which everything is. It liberates and transforms human beings by freeing them from the shackles of their own—personal or cultural—self-centered passions and desires. In order to become the body of God,

St. Paul tells us, we must die unto ourselves so that Christ may live in us. And to achieve Nirvana, Buddhists tell us, we must be attached to nothing except the strange and empty "thatness" (Tathata) of all reality.

The fundamental awareness of this ultimate reality in the various religious traditions is not an explanation or hypothesis, nor is it a mere belief in the existence of a God or First Cause outside and beyond nature, but a *nonconceptual experience* which religious scholars refer to as mystical.

Scholars such as Fritzhoff Shuon and Huston Smith believe that the mystical experience constitutes the fundamental core of the various religious traditions. It is, then, not a philosophical conception or hypothesis, but an *experiential awareness* of and identification with the inexplicable and transcendent *reality* or *actuality* of everything that is, including oneself. That reality—by whatever name—is ineffable, not reducible to any sort of verbal or other representational understanding. Knowledge about it is not the same thing as the reality itself. Whatever knowledge we may have about it is—as contemporary scientists might say—simply a model of it which should never be confused with the reality itself. Furthermore, it is experienced as an interconnected and interdependent whole or one. It is not itself a thing or identical to things, but is perceived inevitably accompanying those things (including the whole set of such things which we call "nature") as their remarkable and miraculous power-to-be. As the medieval mystic Meister Eckhart put it, "God in things is activity, reality and power." Lao Tzu expressed the same thing when he said "from wonder unto wonder existence unfolds."

Not to be experientially aware of this ultimate reality is from this spiritual point of view to exist in a sort of numbed and unawake (slumbering) state. That numbed state is a kind of illusion ("maya" in Hinduism) or ignorance ("avidya" in Buddhism) in which a person loses sight of ultimate reality in favor of dealing with and caring about isolated parts of reality *as if* they were ultimately real. Thus, for the most part we live unaware of the single and mysterious ultimate reality which—again inexplicably—astonishingly happens to be.

Our spiritual traditions have precisely called us to wake up and open ourselves to that wider reality to which Havel refers, and they have developed a variety of tactics and disciplines to facilitate that process, and thus to transform how we actually live. The fundamental moral stance which flows from such a spiritual transformation is that all of life, being in fact a mode and manifestation of that ultimate reality, is holy and thus intrinsically valuable. Love and compassion now are not limited to other human beings, but find their objects in all of nature in so far as it *is*. We are and we do what we most care about.

I have argued elsewhere[4]—as have others—that religion has to do with *interpretive understanding* rather than empirical hypotheses or matters of fact, an interpretive understanding based on direct mystical experience rather than explanatory models and belief systems. That is, religion involves an interpretive understanding of life as a whole and the human role and destiny within it rather than rational and/or scientific *explanations* of things. To quote theologian Langdon Gilkey:

> Religion asks different sorts of questions [than science], questions about meaning. Thus religious myths, symbols, doctrines, or teachings answer these sorts of questions. Why is there anything at all, and why are things as they are? Why am I here, and who am I? Who put me here and for what purpose? What is wrong with everything, and with me? And what can set it right again? What is of real worth? Is there any basis for hope? What ought I to be and do? And where are we all going?...Religion, in other words, tends to answer—or to try to answer—our *ultimate* questions: questions of ultimate origins (where did it all come from?), of ultimate worth (what is the point or meaning, the why of life?), of ultimate destiny (where are we all going?).[5]

These all-inclusive interpretations of what it *means to be* are made available through stories, particularly creation or cosmogonic myths that relate this life to sacred origins or an ultimate order of being. In short, creation myths make available a wider and deeper reality beyond self-centered or anthropocentric concerns, a reality which provides an interpretive understanding of life and how to live in the light of it. In his analysis of the implications for theology of contemporary cosmology, physicist and theologian Ian Barbour writes that

> the function of [religious] creation stories is not primarily to explain events in the distant past, but to locate present human experience in a framework of larger significance. Creation stories manifest the essential structure of reality and our place in it. They provide archetypes of authentic human life in accord with a universal order. They are recalled and celebrated in liturgy and ritual because they tell us who we are and how we can live in a meaningful world.[6]

In fundamental ways myth and science seem to meet different needs. Myth and the spiritual aspect of our lives provides a broader meaning in life, while scientific understanding is about how things work. If scientific hypothesis and explanation are the vehicles for genuine understanding of how nature works and thus for the human need and ability to find some control and security within nature, mythology and stories are the vehicles for spiritual insight and development in human life. The former without the latter leads to a soulless and self-centered form of demoralization; the latter without the former leads on the contrary to a spiritual life of naive magic. As Havel noted, it is the former possibility that seems to prevail in our modern industrial and consumer societies. In our haste to gain security we have sacrificed spiritual vision and connection. We lack an appropriate balance of these two basic needs in ourselves. Our human nature cries out for both security and spiritual depth, not one to the exclusion of the other.

This schizoid breach between science and religion has been a deep and abiding problem in our modern world, influencing how we think about ourselves and how we behave in life. It shows up in the Cartesian dichotomy between matter and spirit (or soul). This in turn has led us to think that we are more than and hierarchically beyond mere matter and nature insofar as "we" are the result of an infusion of "soul" substance at our conception. Nature is turned over to science and religion is left with what remains: the immaterial soul and an increasingly abstract God who is thought of as before and outside nature. The split between science and religion is hardened: Science deals with objective knowledge, and the realm of mere subjective value and meaning is all that is left for religion. Thus our scientific and technological modern culture appears to be nihilistically adrift in a purposeless universe while our religion seems to be a collection of ad hoc claims based on no evidence at all or simply dogmatically asserted to be revealed by God.

As many feminists have observed, however, human nature contains the need for both science and religion, the head and the heart—real understanding that can provide some control and security in a not always hospitable nature as well as a certain reverence and appreciation for life, the male side of life if you will, as well as the female. The balance between these two human needs certainly began to be upset during the development of the early church (if not earlier), a development which emphasized conceptual understanding as true or false "beliefs," an increasingly abstracted and withdrawn Father god, and a hierarchical church organization dominated by men. This imbalance grew deeper with the Enlightenment and modern success in scientifically (and politically) dominating and controlling both nature and nonwestern peoples (through colonization) in the succeeding centuries. We have inherited from this imbalance a kind of cultural schizophrenia in which we find it nearly impossible to live whole lives inclusive of both our heads and our hearts. Willy-nilly we reduce ourselves to one or the other, but hardly ever both. But because of changes in our understanding of the religious side of our lives and parallel changes in our understanding of the scientific side that have accompanied the truly revolutionary developments in cosmology and biology, we may now be on the threshold of an era in which we might right the balance and, here is the hope, become whole again.

We seem to be moving toward such a balance of these needs. Science and religion—albeit not identical endeavors—are no longer enemies of one another, nor even incompatible. They seem in fact to have entered into a profound and helpful conversation—something our European tradition has not seen for three hundred years.

Religious Symbols

Because myths are made up of symbols and indeed function as symbols, we need to say a few words about religious symbols before continuing our exploration of myth.

Symbols are necessary and universal aspects of spiritual life. I don't mean by symbol, here, something which is *merely* symbolic—that is, "not real." If it is alive in a human culture, a religious symbol doesn't represent or stand-in for something else, as the word "God" is so often taken to *name* and *represent* some other (than the word) reality. Rather, living symbols are like windows through which humans encounter what is for them ultimately most real and meaningful about life. I am claiming that the sacred in various traditions is encountered always and only through religious symbols.

Such religious symbols can be sounds, words, things, images, moods, metaphors, or persons. The White Mountains of New Hampshire or the Grand Canyon in Arizona are of course highly symbolic for many people, material symbols through which people encounter what is ultimately holy and meaningful for them. Clearly, cities such as Jerusalem, Mecca, or Rome are symbolic for millions. The river Ganges is deeply symbolic for Hindus. Waves on the ocean become symbols for certain Mahayana Buddhists, as does the yogic mantra, Om Mane Padme Hom. Indeed, yoga is a behavior which is highly symbolic, especially in its highest stages. "Reason" is surely symbolic of what is ultimately powerful and significant about human life for many humanists and rationalists. And, of course, Jesus is symbolic for Christians in so far as they see in and through his life what is ultimately meaningful about human life and destiny. Dutch theologian and New Testament scholar Edward Schillebeeckx emphasizes this symbolic nature of Jesus as the Christ by calling him the "face of God." Symbols of various kinds, then, are the means through which human individuals and cultures discover or encounter and take on an ultimate interpretive understanding of life.[7]

Myths themselves, and particularly creation myths, are made up of symbols and are themselves—if they are still vital—symbolic in this sense. They are symbolic windows or narratives which disclose or make available an ultimate vision of life as a meaningful whole and our role and destiny within it.

Creation Mythology

"Mythology" comes from the Greek word "muthos" which means a story or something told or said. A story involves a series of actions which, like the notes in a melody, are individually significant only in so far as they are interrelated parts of a meaningful whole. The plot makes the individual actions or events more than a mere chronicle by tying them together into a particular meaningful whole. Narrative, then, is especially suited for providing an all-inclusive sense of religious significance because it permits us to see the diachronic events involved in creation as parts of a synchronic, meaningful whole. In other words, creation myths afford a vision of the parts of creation by linking them to a narrated whole that includes them.

Myths are stories about the sacred and the relationship of the world and human beings to it. Mythology bifurcates reality into two levels: a transcendent or deep level of meaning (heaven, the abode of the gods, sacred reality) and the separated, dependent, lesser, and ordinary world of nature and the human community as it is presently constituted. As David Klem notes in his 1986 book,

> Sacred myths speak of the acts of divine beings in setting the goals for human beings, the meaning of human suffering and trials, and the sequence of life

stages through which every individual must pass. Myths intend the integration of individual and collective life within the sacred order of being. Individuals and cults internalize the mythological narrative, allowing it to shape their lives…. Ritual reenactment of myth ensures the public, social status of the myth and enables the internalization of meaning. Because of public ritual, myths are not just stories, but are scripts for performance.[8]

The divine level is considered "real," indeed "reality" itself. It is perfect, eternal, stable and unchanging, holy, and of fundamental and ultimate significance; the ordinary world is as it were less real, imperfect, temporal, and changing, and dependent upon the Sacred for both its existence and whatever meaning, order, and success is achieved within it. Mythology—creation mythology in particular—discloses a human awareness of a transcendental reality beyond this world, but reflected within it, what Emerson called "a world elsewhere." Joseph Epes Brown,the anthropologist and friend of the Lakota holy man, Black Elk, put it this way:

> It is often difficult for those who look on the tradition of the Red Man from the outside or through the "educated" mind to understand that no object is what it appears to be, but is simply the pale shadow of a Reality. It is for this reason that every created object is wakan, holy, and has a power according to the loftiness of the spiritual reality it reflects. The Indian humbles himself before the whole of creation because all visible things were created before him and, being older than he, deserve respect.[9]

The writer T. C. McCluhan describes the Australian aboriginal notion of Dreaming this way:

> The Dreaming is the other world. It is an everlasting and hallowed world that is peopled with great mythic spirit beings. "It is a big thing; you never let it go…." It is

"like engine, like power, plenty of power; it does hard work; it pushes," explained one Aboriginal. The Dreaming gives meaning to life, bestowing upon it depth and resonance through memory. The Dreaming is the ground of being. It is also known as the Law: the generative principles of past, present and future; the body of ethics and the code of life. It has been called the "plan of life." In other words, The Dreaming gave order to the world and laid down the Way (of the ancestors) for humans.... The Dreaming is the period at the beginning of time when enduring shapes took form, enduring connections were established, and enduring events and exploits happened.[10]

We should note, however, that although it is sometimes pictured as such this transcendent reality or "world" is not literally located somewhere any more than are Plato's ideas. If it were, it would be finite since only finite things can be spatially located. Rather, it is transcendent in that it is metaphysically *other and more* than this finite world in which we find ourselves, not reducible to it. It is a deep structure or metaphysical order beneath or within nature, always accompanying and apprehended with finite things, but itself not one of them. It is nonfinite (infinite), then, and as such is thought to be the power, source and ground of being of all the finite creation that is. It is religiously experienced in wonder and awe as the mystery or miracle of being, not this or that finite entity, but the strange and wonderful power of things to be, the mysterious (inexplicable) fact that anything is at all. The well-known theologian Leonardo Boff declares that such mystery

is not an enigma which, once explained, disappears. Mystery is the dimension of depth to be found in every person, in every creature, and in reality as a whole; it has a necessarily unfathomable, that is inexplicable aspect.[11]

In his history of Gnosticism, G. Filoramo provides an interesting example of this sort of bifurcation of reality in a creation myth of origins. The fundamental myth of Gnosticism was to trace back the emergence of this life of embodied and thus "fallen" existence to the divine "light" which is its origin and which constitutes its "pleroma" or fundamental and divine substance. The point of such a creation story is not disinterested contemplation, but

> immersion in the vital, throbbing reality of origins, the ability to tune into the divine energy, to allow oneself to be penetrated by it to the point where one is possessed and transformed by it.... Myth thus acquires the function of salvation. It describes the way of salvation, reminding the Gnostic of his true origins and showing him how to escape from the cosmos. But above all, like all myth, that of the Gnostics is essentially a story of origins: there lies the key of all that one thinks one possesses. But the "origins" of the cosmos coincide with the pouring forth of Being.[12]

In the words of the ancient Gnostic Theodatus, the fundamental knowledge (gnosis) reveals "who we are, what we have become, where we have been cast out of, where we are bound for, what we have been purified of, what generation and regeneration are."[13]

Another typical creation story is the Huai-Nan Tzu from the Han period (206 BCE–220 CE) in China. It is a mixed Taoist and Confucian story. Notice the way in which the story discloses how multiplicity and difference (heaven and earth, yang and yin) emerge from a primal unity, and that the purpose of the story is to connect "the true man" with the original and formless "Great Beginning."

> Before heaven and earth had taken form all was vague and amorphous. Therefore it was called the Great Beginning. The Great Beginning produced emptiness and emptiness produced the universe. The universe produced material-force which had limits. That which was clear and light drifted up to become heaven,

while that which was heavy and turbid solidified to become earth. It was very easy for the pure, fine material to come together but extremely difficult for the heavy, turbid material to solidify. Therefore heaven was completed first and earth assumed shape after. The combined essences of heaven and earth became the yin and yang, the concentrated essences of the yin and yang became the myriad creatures of the world. After a long time the hot force of the accumulated yang produced fire and the essence of the fire force became the sun; the cold force of accumulated yin became water and the essence of the water force became the moon. The essence of the excess force of the sun and moon became the stars and planets. Heaven received the sun, moon, and stars while earth received water and soil.

When heaven and earth were joined in emptiness and all was unwrought simplicity, then without having been created, things came into being. This was the Great Oneness. All things issued from this oneness but all became different, being divided into the various species of fish, birds, and beasts.... Therefore while a thing moves it is called living, and when it dies it is said to be exhausted. All are creatures. They are not the uncreated creator of things, for the creator of things is not among things. If we examine the Great Beginning of antiquity we find that man was born out of nonbeing to assume form in being. Having form, he is governed by things. But he who can return to that from which he was born and become as though formless is called a "true man." The true man is he who has never become separated from the Great Oneness.[14]

Of course, the Genesis account of creation similarly traces back (and thus accounts for) the multiplicity and different forms of creation to a single and ultimate unity, God. In Genesis 1:1-

23 we have the first version of the story, that of the so-called Priestly ("P") source. In this telling, the earth was initially formless and void, and God goes through the seven days of creation by naming different aspects of creation, starting with light and darkness (opposites like yin and yang). In the second version (Genesis 2:4-23) which has been edited together with the earlier story, God is like a potter who first fashions Adam from the dust of the ground and breathes life into his nostrils, and then goes on to place him in the Garden of Eden. This is the Jahwistic ("J") narrative source, and the rest of Genesis through book 11 simply recounts the emergence and chronological development of human life. Thus, "J" is accounting for both the emergence of a multiplicity of things on the earth from God as well as the shadowed aspects of this life such as a sense of estrangement from God, alienation between humans, the host of languages which divide and separate us, and so on. The Priestly Source, thought to have been written in the fourth century BCE, integrates this narrative into the later Jewish epic recounted in Exodus, Leviticus, Numbers, Deuteronomy, and the chronicles of Samuel and Kings in order to found the second temple and its Priestly school upon the divine origin and history spelled out in Genesis. In other words, the Second Temple and its priesthood constitute the sacred foundation of post-exilic Israel, in effect thought to have been established since creation on the ultimate and fundamental reality and authority of God.

The Code of Hammurabi makes the similar point that the great king—with the aide of Marduk—founded his temple culture in Babylon at the time of origins. This tying of the culture to the original gods is common and many scholars think that it represents an effort to project the cultural order back to the divine origins. New Testament scholar Burton Mack points out that the motivation for doing this was certainly not a trivial one for "it made them right, legitimate, centered, and at home in the world."[15]

Again, the point of these creation stories is to connect this life with the fundamental, ultimately real divine origin, thereby interpretively seeing and understanding human life, our sense of estrangement, and especially our need for salvation, in the light of that origin.

> In the myths of creation of the cosmos, religious tra-
> ditions express their understanding of the ultimate
> meaning of the world and of human existence. They
> tell of the role of gods and goddesses in creating and
> sometimes even dying through sacrificial dismember-
> ment to make the entire world holy. They describe
> how the world (oceans, land, mountains) came to be.
> And they separate the reality into different realms.[16]

The ancient Chinese creation story of P'an Ku is just such a myth that connects the heavens with the earth and which, through the sacrifice of P'an Ku, sacralizes the earth.

Initially, all there was was a chaos in the form of an egg. P'an Ku was born from that egg, and for 18,000 years (that is, for a long time) he grew, pushing heaven which was considered light up and earth which was thought to be heavy down, there-by separating reality into two planes. The light was yang (male) and the heavy was yin (female). Thus, a single reality (the egg) became differentiated into two, and from such a differentiation ultimately the entire earth emerged. When heaven and earth are fixed, P'an Ku dies and sacrifices parts of his body to make up the earth: his head becomes mountains, his breath clouds, his voice thunder, and his arms and legs the four quarters of the uni-verse. Thus, the story outlines how the diverse life we know emerges from the ultimate unity and how it actually mirrors that unity.

Creation mythologies, then, manifest a haunting awareness of transcendent, ultimate, originating reality reflected in the less-er (because merely reflected) visible universe. This world is a "sacred cosmos"—a meaningful and ordered universe—to the degree that it reflects that deeper, sacred reality. The creation sto-ries narrate how an original one creatively evolves into the man-ifold of diversity and fecundity that characterizes present life.

The Grammar of Interpretive Understanding

I like to call this bifurcation into two levels the *grammar of human interpretive understanding and meaning*. "Grammar," of course, refers to the formal rules of a language whereby the various parts of speech are arranged in order to constitute meaningful discourse. In a parallel way, the grammar of interpretive understanding comprises the rules or structure of myths that permit them narratively to disclose an ultimate all-inclusive meaning to life as a whole. The grammar consists of the bifurcation of reality into two levels, and it is through this double structure that human beings frame ordinary life with an interpretive vision of what life is all about and thereby construct the various human worlds or cultures in which they actually live. The grammar functions by helping us to "see" that ordinary life and world "as" a dependent reflection of the sacred other. The "as" constitutes the interpretive understanding of nature and our lives "seen" as a meaningful whole.

Essential to this process of "seeing...as" is metaphor, or rather, as we shall see, double metaphor: As linguistic philosopher Max Black has put it, metaphor involves using a conventional image drawn from ordinary life as a screen through which to see another.[17] Thus, in metaphor we apply one aspect or characteristic of experience to another on the grounds of shared similarity in order to gain a fuller understanding of the meaning of the latter. For example, to say, as does Plato, that conceptual understanding is a kind of "seeing" by the mind uses an ordinary form of bodily perception to illumine and comprehend the analogous experience of *conceptual understanding*.

In his *Report to Greco*, Nikos Kazantzakis, known also for his novel *The Last Temptation of Christ*, makes available to the reader an interpretive understanding of God and life as a whole (including human life) by, first, picturing God metaphorically as a merciless and demanding Cry and, secondly, understanding human life in the light of that cry as a painful response and emergence.

Blowing through heaven and earth, and in our hearts and the heart of every living thing, is a gigantic breath—a great Cry—which we call God. Plant life wished to continue its motionless sleep next to stagnant water, but the Cry leaped up within it and violently shook its roots: "Away, let go of the earth, walk!" Had the tree been able to think and judge, it would have cried, "I don't want to. What are you urging me to do! You are demanding the impossible!" But the Cry, without pity, kept shaking its roots and shouting, "Away, let go of the earth, walk!"

It shouted in this way for thousands of eons; and lo! as a result of desire and struggle, life escaped the motionless tree and was liberated.

Animals appeared—worms—making themselves at home in water and mud. "We're just fine here," they said. "We have peace and security; we're not budging!"

But the terrible Cry hammered itself pitilessly into their loins. "Leave the mud, stand up, give birth to your betters!"

"We don't want to! We can't!"

And lo! after thousands of eons man emerged, trembling on his still unsolid legs.

The human being is a centaur; his equine hoofs are planted in the ground, but his body from breast to head is worked on and tormented by the merciless Cry. He has been fighting again for thousands of eons, to draw himself, like a sword, out of his animalistic scabbard. He is also fighting—this is his new struggle—to draw himself out of his human scabbard. Man calls in despair, "Where can I go? I have reached the pinnacle. Beyond is the abyss." And the Cry answers, "I am beyond. Stand up!" All things are centaurs. If this were not the case, the world would rot into inertness and sterility.[18]

We have a double metaphor here through which we catch a glimpse of what is ultimately meaningful about life and of our own role and destiny in the light of it. By *seeing* God *as* a primordial and demanding Cry, we come to *see* life as a whole and our own human lives in particular *as* a struggle and response to that Cry. It is this "seeing" and then "seeing…as" which constitutes the explicit interpretive understanding Kazantzakis is making available to us in the story. Notice, by the way, that the double metaphors are related through a story (remember, myths are stories)!

There are, of course, numerous other examples of this grammar of interpretive understanding in the world's long list of creation myths. We often see God (or goddess) pictured as a fertile mother earth. And thus human life is seen over against that as a kind of fecund birthing. In the Bible, God is metaphorically "seen" in a variety of different ways: as lover, as shepherd, as conquering general, as sovereign king, or as "daddy" (Abba) for Jesus. Each of these images is a metaphor which then leads to seeing our ordinary, everyday lives metaphorically as like a loved one, a senseless but obedient sheep which needs protection, as a loyal footsoldier in God's army, as a loyal subject in God's Kingdom, or as a beloved child of his caring father. The familiar image of Christ Pantokrator expressed in the mosaics of fifth-, sixth-, and seventh-century romanesque churches is another such metaphor. Just as Caesar is the distant but all-powerful ruler of the Roman Empire who rules with his council of landed gentry (lords), so the metaphor asserts that Christ is a sort of cosmic Caesar who, with his council of saints, rules the entire universe. This life, then, is interpretively "seen…as" a sort of patient serfdom, not unlike that of the serfs who served on their lord's estate and whose lives were shaped by their lord and transcendent Caesar. Various metaphorical images of ultimate reality thus lead to the variety of interpretive understandings of covenant and faith within the biblical traditions.

But, beyond the metaphors, what *is* this sacred reality? We cannot directly answer that question. We can say, however, that it is not a finite entity of any kind, which is to say that it is not an "it." As such it is not a determinate something, but a transcendent and indefinable, ultimate ground of being, a power of all things to actually be which can only be experienced mystically and expressed symbolically and metaphorically. It is what is ultimately real, the ground and foundation of all that is, indeed *reality as such* rather than explanatory cognitive models of it. For human beings who encounter it and identify with it in the mystical experience of wonder and awe, it is that which is most meaningful about life and the focus for religious yearning, discipline, and destiny.

Through creation stories, we see that the eternal sacred breaks into ordinary space and time to found and permeate human worlds with meaningfulness. Cultural anthropologist Clifford Geertz states this classically in his influential essay, "Religion as a Cultural System."

> [S]acred symbols function to synthesize a people's ethos—the tone, character, and quality of their life, its moral and esthetic style and mood—and their world view—the picture they have of the way things in sheer actuality are, their most comprehensive ideas of order.[19]

Human cultures are symbolic in so far as they are founded upon and embody a systematically interrelated set of ideas about the meaning of life. By telling the story of the relationship of the ordinary and the sacred level beyond it—and by actually performing the story in their public rituals—human communities transform a mute and meaningless nature into a symbolic, human world that is meaningful in so far as it is understood to be derived from and reflective of an absolute reality beyond it. Such myths, then, lay out the divine origins of this life, explain or otherwise rationalize the existence of suffering and evil by placing it within a more inclusive story, and finally provide an all-inclusive vision of the human role which has evolved out of that original and ultimate reality.

The great historian of religion, Mircea Eliade, claimed that for primal cultures, creation mythology recites how nature and the tribe come to be,

> how something was accomplished, began to be. It is for this reason that myth is bound up with ontology; it speaks only of realities.... Obviously these realities are sacred realities, for it is the *sacred* that is preeminently the *real*.[20]

Creation myths, then, narratively connect us to a wider, significant reality to which we belong. They are not hypotheses at all, but interpretive understandings of our lives in the light of that wider order that inform us about how to live meaningfully. The truth they seek in this process involves the existential question of *how to live* fully and deeply rather than the kind of hypothetical truths involved in scientific understanding and explanation of nature and ourselves as entities who have evolved from it. This interpretive understanding of life entails a religious imperative for a human community to live focussed upon that ultimate reality or sacred ground of being from which the entire manifold of *what is* has evolved and that as such is fundamentally holy and meaningful. Such myths, then, contain an implicit demand for people to shift their attention and lives from the unreal to the Real, from self-centeredness in the here and now to a centering on the eternal order of being that underlies and encompasses all that is: to actually *live* differently and thereby become their true selves. People seek to walk a sacred path in life with their minds and hearts fixed on that sacred power-to-be. In a very real sense, mythology shapes and structures people's everyday behavior by helping them to notice the difference between what their ordinary lives are like and what they *might* or *ought* to be like, thereby enabling them to interpretively understand their lives in the light of that spiritual imperative.

Above all, these founding myths are all-inclusive and synthesizing narrative interpretations of what it means to be, ways of seeing our human lives and the goals we seek to achieve within the encompassing and ultimately meaningful broader reality of which we are a part. As stories, creation myths are uniquely qualified to provide such an inclusive understanding precisely because they are able to stitch together episodic events into a single meaningful whole. We tell stories to feel at home in the universe. Such mythology is essential to human life for it is the way we answer the questions we have always asked nearly fresh from the womb: "What's going on?" "Where did we come from?" "What are we doing here?" "Where are we headed?" "How ought we to live?"

Traditional creation mythology, then, is a story about the whole of reality that: (1) manifests and makes available to human consciousness a wider and deeper reality than our ordinary reality; (2) discloses that all of reality is a single, meaningful and inclusive whole from which all the different aspects of the cosmos are derived (it is both one and many); (3) manifests the worthiness and intrinsic value of that wider reality in so far as it is seen to be fundamental, ultimate—that without which the dependent aspects of nature would neither be nor be as they are; (4) shows precisely how all of nature is dependently derived from that one; (5) divulges our rootedness and connectedness to the larger life to which we belong by showing specifically how we belong to it and by showing what our role and destiny is within it—i.e., shows us how we fit into life; (6) stirs feelings of reverence and awe by inducing a sense of wonder;[21] (7) stimulates a sense of gratitude not only for the seemingly gratuitous gift of life but for being aware of it and the wider reality to which we belong; (8) teaches us to be more humble and less self-centered in the face of such an immense reality; and finally, (9) transforms the lives of those who are touched by the story by inducing them to live in the light of the ultimate reality it narratively makes available. Myth, especially creation myth, is a script for ritual performance in which the participants internalize the meaning of life that the myth narratively discloses and are thereby transformed. In short, creation mythology induces in those for whom the myth is vital and alive a deep sense of reverence for a holy reality which grounds and sustains both nature and human culture.

What we need to do now is to outline the story of our universe as it is contained in the new, scientific cosmology that has emerged over the past fifty years. My contention here is that this cosmology is not only a scientific understanding of our universe but also a religious creation story which, like all creation stories, displays the unfolding of a great mysterious reality with immense significance for our lives today.

It's all a question of story. We are in trouble just now because we do not have a good story. We are in between stories. The old story, the account of how the world came to be and how we fit into it, is no longer effective. Yet we have not learned the new story.... A radical reassessment of the human situation is needed, especially concerning those basic values that give to life some satisfactory meaning. We need something that will supply in our times what was supplied formerly by our traditional religious story. If we are to achieve this purpose, we must begin where everything begins in human affairs—with the basic story, our narrative of how things came to be, how they came to be as they are, and how the future can be given some satisfying direction. We need a story that will educate us, a story that will heal, guide, and discipline us.... No community can exist without a unifying story.

§ Thomas Berry, The Dream of the Earth

Science in our new situation in no way argues against the existence of God, or Being, and profoundly augments the sense of the cosmos as a single significant whole.

§ Menas Kafatos and Robert Nadeau,
The Conscious Universe

The New Cosmology

Cosmology" means a theory of the universe as a whole. Since that universe is an emerging reality—one in process—I use the word to mean an understanding of both its origins and its evolution to the present day. In the past fifty years, a new scientific understanding of the entire universe has emerged, an understanding that is fundamentally a *story* of the whole twelve-to fifteen-billion-year emergence of nature as a whole, including most recently ourselves. That story is both a scientific cosmology and a religious cosmogony, both scientifically true and religiously revelatory and meaningful. Because it is scientific and religious, it may help us (as cultural historian and theologian Thomas Berry and others claim) to reorient ourselves within the whole and thereby provide a comprehensive guide for living in our time.

A Shift in View

Before outlining that story, I'd like to discuss the social and intellectual context in which it has appeared, for it entails a massive cultural shift in how we look at nature and how we understand our place within it. In other words, we need to step back first of all from the details of that scientific narrative of creation in order to grasp its larger significance.

Just think of the remarkable changes in how we see the universe and our place within it brought about by the scientific and technological revolution begun in the Enlightenment. Such a sea change in our understanding of the reality in which we find ourselves has begun to permeate and shape our spiritual and moral values and behavior. After all, as philosopher Holmes Rolston has written,

> we always shape our values in significant measure in accord with our notion of the kind of universe we live in. What we believe about the nature of nature, how we evaluate nature drives our sense of duty.[1]

Since the Copernican revolution, science has revealed a universe in which the earth is not at the center of our solar system and in which that solar system is certainly not at the center of the universe. It is of course only in this century that science grasped the notion of "galaxies." In the early 1920s, astronomers observed that our solar system was part of the Milky Way galaxy, which contains billions of suns, and that there are two nearby clusters of stars. It wasn't until stronger astronomical instruments were developed, however, that they came to realize that our galaxy is but one of about fifty billion galaxies, each with billions upon billions of suns. Then, in 1927, Edwin Hubble discovered the so-called "red shift" which indicated that all of these galaxies are receding from each other at a rate which increases with the distance of each from the others—the more distant the faster they are moving apart (Hubble's constant). Only in this century, then, has the true immensity of the universe worked its way into our consciousness.

Now, add to that our altered understanding of both the age and the nature of that immense universe. Until rather recently, creation was quaintly thought to have occurred about eight or ten thousand years ago. Our understanding of that creation was not only that it was recent and centered on the earth or our solar system, but also that it was fixed and static. The geological forms, animals, and flora were given as we know them now, the mountains and deserts were set, and the oceans and continents were established in their present form and pattern. But starting in the eighteenth and nineteenth centuries, science revealed ancient geological strata and fossils of no longer existing creatures. It became clear that the very landscape itself, far from being recent and fixed in nature, is both older than ten thousand years and (more surprisingly) has been changing and evolving over immense eons of time. Then Darwin came along and made us aware of biological evolution. The various species of life are not only not as recent as had been thought, but have themselves evolved from earlier forms. And in the twentieth century, of course, we have come to realize that the very oceans and continents themselves are not fixed but fluidly emergent and continuously in motion.

By the early years of this century, scientists estimated that the earth was at least 200,000 years old. Finally in 1968 Robert Wilson and Arno Penzias of the Bell Labs in New Jersey stumbled upon an astonishing discovery. They determined the temperature of the background radiation left over from the so-called "big bang." Astonishingly, we have a remnant of that original creation event right here in the present—and all around us. Although this measurement in itself does not establish the age of the universe, extrapolating from the rate of expansion of the galaxies has led scientists in recent years to push back creation in time to between twelve and fifteen billion years ago.[2]

Thus, far from inhabiting a recently created, static and fixed cosmos centered on the earth, we have come to realize that an awesome immensity of time was involved in the creation of our universe—five or six billion years for the earth and twelve to fifteen billion for the universe as a whole. At the same time, we have come to understand the breathtaking size of that cosmos. And, finally, we now realize that the nature of creation is precisely (and through and through) one of flux and emergence. This tremendous shift in how we see things, taking place in such a short time, has been a shock to our culture and may ultimately lead to significant spiritual and moral transformation and development.

Simply put, this cosmology understands nature as a historical or temporal unfolding over billions of years, a story which entails a vision of the whole of the universe as seen through all the sciences. Rather than describing nature as a complicated machine, as had been the case in earlier scientific understandings, this cosmology pictures it more as a living, organic process. This shift in our perception of the universe forces us to revise our understanding of God from that of an initial clockmaker-designer standing apart from the machine to something more like an immanent force or power-to-be which unfolds from the original singularity into the astonishing, interrelated, and ordered manifold of reality we have come to recognize. Biologist Rupert Sheldrake spells this out.

Once again it makes sense to think of nature as alive. The old cosmology of the world-machine, with the divine engineer as an optional extra, has now been superseded within science itself. This completely alters the context in which the relationship between God and nature can be conceived. For if the entire cosmos is more like a developing organism than an eternal machine, then the God of the world-machine is simply out of date.[3]

Changes in Science

Before going on to the story of the emerging universe which has made this new worldview possible, we need to consider these questions: Why has such a cosmology not been available until now? What does the very nature of contemporary science contribute to the new cosmology so that it turns out to be both scientific and religious in nature?

A theory of cosmology, or natural theology as it was earlier called—the human attempt to understand or comprehend the natural world in its entirety—has not really been possible since the seventeenth century until very recently. There are at least three reasons for this.

First, the Cartesian worldview pictured reality as divided into two aspects, "mind" and "matter." Theology was limited to the former while mechanics and the new science in general were assigned to the latter. Theology was thus a priori excluded from any commerce with nature: Nature was desacralized and a natural theology or cosmology was rendered impossible.

Second, by the middle of the nineteenth century the natural sciences had fragmented into a number of separate disciplines, each with rather different foci and forms of discourse. It was the business of those fragmented sciences to concentrate on the *parts* of nature to the exclusion of the whole and the interdependence of those parts. Since theology was ruled out of the nature game, no particular discipline was assigned to explore the whole. Cosmology, then, fell through the cracks because neither theology nor science were in a position to deal with it.

Third, not only was each science insulated from all the others so that a full picture could not emerge, but also the very model of doing science seemed to preclude an overarching picture of the whole in so far as the goal was to study and reduce all such wholes (including "nature") to their parts. The wholes we perceive in our ordinary experience, from this point of view, are really conglomerations of ultimately irreducible atoms. These bits and parts of all things are only externally related to each other and the wholes are heaps or collections of those parts. On this view, of course, there is nothing newly characteristic ("emergent") about such new wholes as molecules, organic cells, or organisms that have evolved from the conjunction of their parts: They can be reduced to and are simply the sum of their parts. In modern physics and biology, that those wholes display novel characteristics and thus are *more* than the sum of their parts was systematically overlooked because of the scientific assumption that the real riches lay in the parts to which, ultimately, everything could be reduced. This reductionistic assumption sometimes leads people to think that such wholes as trees, for example, which appear in our ordinary experience, are in fact less real or perhaps even illusory manifestations of their ultimate parts. Those atomic parts constitute in Platonic fashion (although they are not identical with his "ideas") an *eternal* and *changeless reality*. The reductionistic attitude is thus actually a metaphysical rather than a scientific assumption that what we are scientifically seeking must be eternal and changeless, and anything less than that (like trees and the rest of the modes of nature) are to that degree less real. Wholes—and the whole—was simply overlooked because science was too fascinated with the parts to notice them (and it).

This leads us to our second question. What is it about contemporary science that is leading to the emergence of a view of the whole while at the same time is making room within that cosmology for genuine religious experience? In other words, what is it that is making possible an inclusive scientific cosmology which is at the same time a natural theology? There seem to be at least three reasons for this remarkable development.

First, nineteenth- and early twentieth-century science inherited from the Enlightenment a rationalist assumption that science would ultimately demonstrate that reality is entirely rational and knowable. This rationalist assumption has been brought into question within twentieth-century science itself. Heisenberg's Uncertainty Principle, for example, stipulates that one can learn *either* the precise position of a particle in motion *or* its particular momentum, but *not both*.

> The wave-particle duality of matter leads to an intrinsic uncertainty in nature, that is, an uncertainty not arising from our ignorance or inability to measure but an absolute uncertainty. Nature must be described by probabilities, not by certainties.[4]

Our knowledge of the subatomic world is limited, then, to partial answers which to a large degree depend on the questions we ask.

In developing his quantum mechanics, Niels Bohr further eroded the rationalist a priori embedded in modern science by showing that the observer of subatomic entities, far from being neutral and objective, actually contributes to what is being observed: a wave or a particle. The act of observing, then, disrupts and alters the real situation under observation. (This is the so-called Copenhagen interpretation of Heisenberg's principle.) We can never know nature completely or as it is in itself because, as observers, our senses, minds, and instruments shape and condition whatever we come to know about it.

Finally, in his incompleteness theorem the mathematician Kurt Gödel in 1930 established that the validity of a mathematical system cannot be demonstrated within that system itself. That is, the validity of any mathematical system depends on a frame of reference *beyond* it. It follows that no system can ever hope to fully comprehend or exhaustively explain reality, for it always depends for its validity on a reality beyond it which by definition it does not and cannot comprehend.[5] Gödel's proof was limited to mathematics, but because mathematics is the language of much of science (especially physics) his proof seems to show that science itself can never reach closure. Mathematics and science are both open-ended. As physicists Menas Kafatos and Robert Nadeau write,

> what the theorem reveals in regard to the limits of mathematical language closely parallels in our view what is revealed in our new epistemological situation in a quantum mechanical universe—the universe as a whole, or reality-in-itself, cannot "in principle" be completely disclosed in physical theory.[6]

Human understanding in general and science in particular are in principle limited and necessarily incomplete. We appear to be attending the funeral of rationalism.

Gödel's insight into mathematical systems is mirrored by structuralists and systems theorists who insist that truth and meaning as associated with both things and words are always context-bound and thus can never find rest in a contextless presence. There can be no final and complete understanding and truth, then, because such an understanding, like Gödel's theorem, demands a further or broader context in which it is inscribed in order for it to be meaningful or true. In the words of the postmodern philosopher, Jacques Derrida,

> the play of differences involves syntheses and referrals which prevent from being at any moment or in any way a simple element which is present in and of itself and refers only to itself. Whether in written or spoken discourse, no element can function as a sign without relating to another element which itself is not simply present. This linkage means that each "element" is constituted with reference to the trace in it of the other elements of the system. Nothing, in either the elements or the system, is anywhere ever simply present or absent.[7]

The whole is open-ended and evolving toward new and emergent contexts. There is nothing (with the possible exception of being itself) that isn't a context which both includes subcontexts and is part of larger contexts, and so on without limit. There is, then, no complete rational understanding—no God's-eye point of view which encompasses everything. There are limits to reason where we encounter mystery, not in the form of a gap to be filled, but as the ineluctable mysteriousness of the whole thing.

Adding to this limitation on the reach of our reason is the fundamental shift in the hitherto predominant reductive model of science mentioned earlier. The assumption that wholes (a molecule for example) could and should be reduced to their parts was the unquestioned and entirely pervasive model of explanation in science, whether in chemistry, nuclear physics, or biology and botany. But not only have we not found atoms or particles in terms of which we can explain everything, but biologists such as Ilya Prigogine and Rupert Sheldrake, and systems theorists like Erich Jantsch, have had some success in demonstrating that reality is, on the contrary, emergent. That is to say that the characteristics of wholes cannot in fact be derived from their parts, but are genuinely novel beyond those parts. Of course the various wholes could not and would not exist without their parts. Still, a whole such as a molecule with its particular neutron and electron or wholes such as Eucharyotes, dinosaurs, and mammals—each with unique characteristics—cannot be determined and predicted or predicated on our understanding of their complex parts. Novelty, nonpredictability, and a sort of irrational (not anti-rational) nondeterminism are thus being suggested from within branches of science itself today. The universe is at heart emergent and creative, not reducible at any stage to its parts or to earlier stages that have preceded it. Thus, it is impossible for science to predict the future in any detail, and not just on the level of quantum indeterminacy. Indeed, if the entire universe end to end could be fed into a computer of maximum power and capacity, information theorists tell us that it would be incapable of predicting all future developments. The universe is creatively open.[8] It may be the case, as one version of the Anthropic Principle has it, that creatures such as ourselves could not have come into being without the host of conditions that preceded us. This does not mean, however, that given our understanding of various stages along the way we could have predicted the eventual emergence of Homo Sapiens.

The rationalist dream of an entirely knowable (and thus predictable) universe has been forced to give way to a more modest and pragmatic vision of the scientific endeavor. By showing us the limits of rational explanation in principle as well as in practice, twentieth-century science has in effect reintroduced us to mystery, to the final unknowability of reality as the boundary condition or ultimate frame of reference for scientific understanding. By doing this, science has made room in this universe for mystical experience—the positive experience of the mystery of being rather than simply an unsupported belief we hold until an omnivorous science can find the answer to fill the gap.

Putting this another way, we might say that modern science has become aware of its limits by noting, as did Soren Kierkegaard in the nineteenth century, that our thoughts about reality can never be identical to reality itself. Existence—the fact that something is rather than nothing—is beyond the limits of rational explanation and discourse.

Without such an ultimate framework of mystery, without an awareness of the gulf between reality and thoughts about it, without a sense of the as yet unexplained to draw the thinker on, thinking would halt and the ongoing scientific endeavor would come to a standstill. The new cosmology views science as a process, not a completed task, a process that rests upon and finds its motivation in something else: a spiritual encounter with the mystery of being. Without this mysterious organic development there would be nothing left to explain and the whole immense project of understanding would grind to a halt.[9]

Furthermore, there is no "pure" spectator left in the postmodern scientific and cultural community. Contemporary science is so powerful that it necessarily impacts the nature it observes. This gives rise to deeply significant ethical and religious questions concerning its ends and purposes. Many contemporary scientists are being forced by their own work to see themselves not as constraintless spectators and theorists, standing outside the ethical and spiritual realm as they pursue "pure science," but rather as moral and religious agents and participants who bring to their task interpretive understandings and perspectives on life and whose scientific endeavors have significant practical import both for nature and for the human community. Nuclear weapons, germ warfare and genetic engineering are examples of this spillover of "pure science" into the human and natural communities. In encountering these practical limits, scientists are being forced to come to terms with moral and spiritual issues and ends of life. The power of science and technology is pushing many scientists to take responsibility for their actions as *whole* human beings rather than some sort of transhuman and transhistorical beings standing outside of and beyond the moral and religious questions the rest of humanity faces.

Finally, there now seems in fact to be a science of the whole beyond the fragmented disciplines within it: It is called *ecology*. The word "ecology" comes from the Greek, oikos, which means "house" or "place in which to live." Thus, ecology is the study of organisms within their "home"—the interconnected physical and life-systems which constitute the earth and the cosmos as a whole. Rather than reducing the whole to separate and independent parts which compose it, we have here a recognition of the systemic web of reality in which each of the parts is interdependently connected to all the others and to the whole. This entails, of course, that human cultures and economic and political systems are not separate and independent realities unconnected with the rest of reality, but subsets within the larger environmental web that constitutes nature as a whole. These human subsystems are sustained by the larger nature of which they are a part, and they can and do decay and deteriorate to the extent that they separate themselves from that synergistic and sustaining whole. Human culture is not a supernatural addition to nature, but an emergent and historically conditioned reality within it.

This ecological understanding of the interconnected whole is fundamentally a grasp of reality as a historical or temporal unfolding, an evolutionary story of emergence and integration over billions of years. Theologian John Haught has expressed this very simply.

> Science has increasingly and almost in spite of itself taken on the lineaments of a *story* of the cosmos. The cosmos has itself increasingly become a narrative, a great adventure. Although there have always been mythic and narrative undercurrents in presentations of scientific theory, the past century has brought forth a scientific vision that, starting from the Darwinian story of life on this planet, has moved back in time to embrace the astrophysical origins of the cosmos fifteen or twenty billion years ago. The most expressive metaphor for what science finds in nature today is no longer *law*, but *story*.[10]

Our understanding of nature in its entirety, then, turns out to be a cosmology, a scientific and theological *story* through which we can "see" nature—as well as the human role within it—"as" a meaningful whole. A breathtaking new possibility has unfolded, the possibility of a creation story that may afford our postmodern world a vision of reality as a whole. We have in this cosmology the opportunity for genuine dialogue between religion and science. Of course, the postmodern form of religion in question is not that of dogmatic answers or a "god of the gaps," but that of mystically encountering in wonder the mystery of being. And the contemporary form of science involved is not the enterprise that envisions a complete set of rational explanations of what is, but the human project—based on such wonder and within the encompassing mystery of being—of articulating the remarkable order of being which the universe displays.

The Story of Creation

We need now to tell briefly the story that constitutes the new cosmology. First, note that it is a story. But second, note that this is a cosmology that does not simply describe the beginning of the cosmos as heretofore but encompasses a continuous creation from around twelve or fifteen billion years ago until now. It is, then, a story of a creative *process* in which reality unfolds into the myriad, novel forms that, amazingly, have come into being. Everything has emerged from the initial Singularity: the fundamental laws of nature, the galaxies, our solar system, earth and everything on it, Bangkok and Benares, your local shopping mall, you and I, the music of Mozart, and science itself.

The details of the new cosmology as we presently understand it are not what is of interest to us theologically. Even if many of those details change (and surely they will), I believe that cosmology will still have momentous religious and theological implications and significance. Is the cosmos ten billion years old or fifteen (eight to twelve billion years if the October, 1994 Hubble Space Telescope measurements are confirmed)? Was there an inflationary period at the beginning or not? Did the galaxies emerge earlier or later than we now think, and perhaps in ways we are not yet able to comprehend? However interesting these questions are scientifically, the religious dimension remains: Much as in earlier traditional creation stories, in this cosmology we are viewing reality as a whole and as an inclusive narrative of an evolutionary process in which all the parts are linked and ultimately derived from a singular one. And what shines through that narrative and induces in us a spiritual sense of wonder is the breathtaking creativity it manifests.

In the timeline on the next page, pay particular attention to the right-hand column, to the steps in which new modes of being have creatively emerged in the evolution of the universe. I want to highlight the fecundity and creative thrust of this evolution, the fact that although those new realities could not emerge without the previous stages, they also could not be predicted given those earlier stages. This is to say with systems theorists like Ken Wilbur that reality consists of wholes made up of parts, and in turn become parts within new and emergent wholes. So, for example, various living creatures are made up of cells which are composed of molecules which are made up of atoms which in turn are composed of particles, and so on. The whole could

CREATION TIMELINE

TIME	CREATIVE EMERGENCE

Initial Stage (15 billion years ago)

TIME	CREATIVE EMERGENCE
0 seconds	Infinite Singularity
10^{-43} seconds (temp: 10^{32} K)	Gravitational force separates
10^{-35} secs. (temp: 10^{28} K)	Strong nuclear force separates
10^{-10} secs. (temp: 10^{15} K)	Weak nuclear and Electro-magnetic forces separate
1-3 minutes	Matter emerges in particles along with helium and hydrogen

Formation of the Galaxies

TIME	CREATIVE EMERGENCE
300,000 years after Bang (2000K)	Hydrogen and Helium form lumpy clouds
1–5 billion years after	Around 50 billion galaxies form

Formation of the Solar System and Earth 5–6 billion years ago

TIME	CREATIVE EMERGENCE
9 - 10 billion years after	A sun (Tiamat) in the Orion arm of the Milky Way explodes as a supernova spewing forth heavy elements such as carbon, oxygen, and nitrogen
10 billion years after	Sun, earth and solar system form

The Emergence of Life

TIME	CREATIVE EMERGENCE
11-12 billion years after	First microscopic forms of life:
3-4 billion years ago	DNA, photosynthesis, and sexual regeneration
700 million years ago	Familiar multi-cellular creatures emerge in the sea
550 million years ago	First shell fish appear. The Cambrian explosion of new forms of marine life—including the first vertebrates—which lead to myriad life forms (fish, plants, animals, mammals) which are familiar today
400 million years ago	Life emerges from the sea
235 million years ago	Dinosaurs appear
216 million years ago	First mammals appear
210 million years ago	Breakup of Pangaea and formation of continents
90 million years ago	Flowering plants predominate

The Emergence of Human Life

TIME	CREATIVE EMERGENCE
4.4 million years ago	First hominid ancestor of both humans and apes and chimpanzees
2.8 million years ago	First Humans: Homo Habilis
2.4 - 1.0 million years ago	Humans spread around world: Homo Erectus
200-300,000 years ago	Archaic Homo Sapiens

The Development of Human Culture and Spiritual Vision

TIME	CREATIVE EMERGENCE
40,000 years ago	First cultural remains of modern Homo Sapiens
35,000 years ago	Neanderthals die out
12,000 years ago	Neolithic culture
3,500 years ago	Classical cultures
450 years ago (1543)	Copernicus and modern culture
Now (15 billion years later)	Science sees the whole in the new cosmology

not exist without its parts, but the new whole cannot be predicted from or reduced to its parts. The parts—particles, forces, etc.—are the condition for the possibility of new wholes that could not be without them. Yet, given those elements (and especially due to Quantum unpredictability *in principle* and the *practical* unpredictability connected with sheer complexity revealed in chaos theory), we cannot predict what new wholes will emerge. It may even be that the human species is not the crown of nature as we have so often thought, but a mere step or stage on the way to something surprisingly greater.

Each new whole, then, is an emergent system with characteristics and behavior beyond those of its parts. From the initial Singularity, the whole of creation is nothing but an arrangement of wholes within larger and emergent wholes, and so on—like a vast set of nesting Chinese boxes. Everything, then, is essentially connected to everything else while at the same time (both as species and as individuals) remarkably different and not reducible to its parts. This new cosmology narratively enfolds these novel eruptions into being a single, interrelated plot or network of dependencies. My point is not that God fills the explanatory gaps of emergence, but rather that the sheer fecund eruption into existence of novelty leaves anyone who perceives it in a state of awe and wonder at the mysterious creative force that animates it.

The barest outline of this incredible story will have to suffice here. We can break the story into six steps or stages (see the timeline, opposite). About fifteen billion years ago an astonishing event occurred, an event called the Singularity by scientists. It was remarkable and, as far as we can tell, unique for several reasons. First, there was no "before" it, for space and time emerged at that moment as dimensions of creation, not (as for Newton) dimensions within which creation occurred and in which as it were nature is suspended. Second, we have the emergence of something new—I mean reality or existence itself. In a way that we can only talk about metaphorically, actuality emerged from eternal emptiness and nothing and began the long fifteen billion year process of unfolding into the myriad and fantastic forms of which we are aware. We call this "the Big Bang," but that name conjures up a picture of a sort of explosion outward, as from a grenade. We know from Hubble's Constant and the so-called redshift that from that moment on the universe has been expanding—initially very quickly in what scientists call the inflationary period during which it doubled in size every fraction of a second. This expansion is not so much an explosion from an initial and initiating "bang" out into empty space as it is an expansion of space and time itself, like (as one scientist has put it) raisin bread dough before it is baked rising and taking the raisins with it. In other words, space and time are not static vessels in which the universe is expanding, as Newton envisaged them, but rather are the expanding dimensions of the universe itself.

In that first nanosecond emerged the so-called four fundamental forces of the universe (gravitational force, strong nuclear force, weak nuclear force, and electromagnetic force) along with particles called photons. The whole universe in this first second was an unbelievably hot soup (10^{32}-10^{15}°C) of blinding light (had there been eyes to see it) expanding at an astonishing rate.

Within seconds, the various elementary particles emerged as well as the most prevalent atomic elements in the universe, helium and hydrogen. In 1965, when Penzias and Wilson stumbled upon the radiation left over from this initial bang, measured at a current temperature of 2.8 Kelvin, this confirmed the existence and subsequent expansion, evolution, and cooling down of the initial Singularity.

Somewhere around 300,000 years after the Big Bang, the expansion of the initial particle soup had cooled down sufficiently (2000°) for the initial particles and helium and hydrogen shaped by the four fundamental forces to form more or less dense clouds of matter, and, because of their combination, to render the universe transparent for the first time. The results of the COBE satellite experiment announced in April of 1992 showed that in fact the background radiation was not a uniform 2.8K, but varied in spots as more or less hot. In other words, it showed that the radiation at this stage was "lumpy." If this had not been the case, then scientists would have remained baffled as to how in the next stage of creation the hydrogen and helium combined to form groups of suns (galaxies) that are not uniformly distributed across the universe. Steven Hawking, among others, called the measurement of the variation in temperature of the background radiation "the most important scientific discovery of our time, if not all time," for it seemed both to account for how we presently find the galaxies strewn about and to confirm the theory (or story) of an unfolding universe which had been proposed earlier.

From about one to five billion years after the initial Singularity, the fifty billion galaxies we estimate today emerged as conglomerations of incredible numbers of stars. These stars and galaxies were something entirely new—they had not existed before, but evolved from the universe as it had up till then formed itself. The lumpy clouds of helium and hydrogen came together to form nuclear furnaces we know as suns. From these stars came the new so-called heavier elements such as carbon, oxygen, nitrogen all the way up to iron. These heavy elements were spewed forth from an exploding and burned-out star (a supernova) and led to a second generation of stars and solar systems.

Between five and six billion years ago (nine to ten billion years after the Singularity), a supernova named "Tiamat" by scientists exploded in the Orion arm of our Milky Way galaxy, blasting a dust of heavy elements into the region. The debris and the heavy elements that it contained as well as the shock wave of the exploding supernova triggered the nuclear ignition of our sun and our solar system was formed along with it. Thus, the earth itself probably emerged around five billion years ago and began a unique process (unique at least to our solar system) of evolution. We can only briefly outline that geological, atmospheric, life, and cultural process of continuous evolution; the sheer creativity and immensity of the process is so rich that the details constitute the complex and interwoven substance of all the sciences, social sciences, arts, and humanities now available to us.

First, the earth needed time to cool down and become more stable. It took almost a billion years for the conditions to become ripe for the emergence of something brand new in the universe—life itself. But in fact it is astonishing how quickly it occurred. Many scientists now think that it was probably within one billion years (or very close to it)—that is, four billion years ago—that the first forms of microscopic life emerged with the prokaryotes. These single-celled algae and bacteria that lived off the chemicals in the ocean regenerated themselves through cell-division, and (for the first time) developed DNA. This meant not only that this first life had a memory but that all of the myriad forms of life (plant and animal) which were to evolve from them could do so because of this genetic structure.

By about 3.9 billion years ago, a descendent of the initial prokaryotes called promethio, had developed something new—photosynthesis. For the first time, the energy of the sun was used to sustain life. These life forms were so efficient at separating carbon dioxide into its constituents of carbon and oxygen that they changed the composition of the atmosphere to about what it is now.

By 2.5 billion years ago, another descendent, by the name of prospero, developed the ability to survive the high percentage of oxygen in the air (about 21 percent) through respiration. And around 2 billion years ago, creatures called eukaryotes developed in which a formed nucleus was separate from the rest of the cell.

Between 700 to 500 million years ago, multicellular creatures and fish emerged in the sea, and sexual reproduction, which supports rapid regeneration and thus an explosion of evolution, was—so to speak—invented. The first shellfish appeared and the Cambrian explosion of marine life-forms, especially vertebrates, took place. This in turn led to the various life-forms of fish, plants, and animals which are so familiar today.

By 400 million years ago, life had emerged from the sea. By 235 million years ago, we find the first dinosaurs, and mammals soon follow. Around 210 million years ago, we have the breakup of the original continent (Pangaea) and the beginning of the formation of the continents as we know them today. About 90 million years ago, flowering plants came to predominate and spread around the globe.

In the ongoing and now speeded-up evolution of new forms of life from the original Singularity, true monkeys and apes had emerged by 90 million years ago. As has been recently discovered, the first hominid ancestor of both humans and chimpanzees developed in Africa around 4 million years ago. Between 2.8 million and circa 300,000 years ago the human species developed from Homo Habilis to Homo Erectus to archaic Homo Sapiens.

All of this evolution involved a creative thrust toward novelty and diversity of forms, albeit within a single whole and with formerly emergent characteristics such as DNA and sexual regeneration intact and evolving. But now, with the appearance of Homo Sapiens, something brand new had unfolded—namely, the development of human culture and the various forms of human arts and religion having to do with meaning and attitude toward life.

By about 40,000 BCE, we have the first physical signs of cultural remains from Homo Sapiens. In the Bear Cave burials of the Alps, we find evidence of a cultural awareness and ritual practice in the face of death, probably indicative of an understanding of a reality beyond death. Thus, humans not only emerged biologically, but developed the whole rich texture of human culture and meaning, what we can call a spiritual or religious attitude toward life. This attitude and accompanying ritual practices continued up through the Neolithic cultures with, some scholars think, their mostly female and agricultural deities and images. The world religious traditions that we are familiar with today emerged in and through the classical nation-state cultures. Finally, modern science and the industrial revolution which it spawned developed in the sixteenth, seventeenth, and eighteenth centuries, and of course continued to grow and transform "modern" culture until now.

Some fifteen billion years after the Singularity, it seems that the evolution of human culture continues and that through it science is becoming increasingly aware of the vast story of creation and of the human place and role within it. In another sense, of course, we might say that the universe has become aware of itself through this biological and cultural evolution of scientific sense and sensibility. Human culture continues to unfold in the symbolic awareness within science and other forms of human understanding of the mysterious existence and emanation of the universe in all of its modes and manifestations. It is a breathtaking achievement, to say the least.

As I have tried to emphasize by stressing the novelty and unpredictability of each stage and phase of the evolving universe, the most startling characteristic is its sheer creativity. Nobel Prize-winning biologist Ilya Prigogine and biologist Isabelle Stengers have stated this simply and elegantly:

> [N]ature is change, the continual elaboration of the new, a totality being created in an essentially open process of development without any preestablished model.[11]

From the initial Singularity and the emergence of the four fundamental physical forces, to the emergence of helium and hydrogen, to the development of nuclear star-furnaces and thus the billions of galaxies that exist, to the creation of our solar system triggered by a supernova explosion some five and half billion years ago, to the emergence of life on earth with the prokaryotes, to the evolution of that life up to the recent cultural phase of Homo Sapiens, to the remarkable unfolding of human culture and consciousness up until our present scientific awareness of the significant cosmic whole which is our ultimate home—it is the sheer creativity of that reality that leaves us breathless.

The generative creativity of this universe is not just found at its beginning nor in each stage and step of its evolution into new modes and forms of reality. It is also manifest at every moment as what is emerges from what physicists call the quantum vacuum. This vacuum can be defined as that from which all forms of reality (particles, forces, and so forth) have been evacuated. It has been shown experimentally to be the nothing or emptiness from which quantum particles emerge and then are reabsorbed back into the background. So, for example, during the Planck era (10^{-43} seconds and earlier) various particles seethe into existence and are immediately annihilated by their antimatter twins. As physicist John Gribben says,

> [T]he quantum vacuum is a seething froth of particles, constantly appearing and disappearing, and giving "nothing at all" a rich quantum structure. The rapidly appearing and disappearing particles are known as virtual particles, and are said to be produced by quantum fluctuations of the vacuum.[12]

This is not a Newtonian bowl-like space before creation in which creation occurs, but the nothing between what is and out of which all reality emerges. It is the vast nothing in deep space between galaxies, the nothing between the proton and the electron which hovers about it in a molecule, the nothing between the cells and synapses of our bodies. This power to be in the face of nothing is in a profound metaphorical sense the face of God or God brooding over the abysmal deep, the remarkable eruption of transcendent being at every moment and in every place in the universe. Such a god is not an entity who throws the switch of creation, but the amazing fact that something actually seethes into being from nothing.

It's not just that we are aware that there is something rather than nothing, then; it is also the astonishing fecundity, diversity, and immensity of that "something." The story reveals the immense variety of being, the emergence of new and unpredictable phases and stages that keep popping into being. It is this generative capacity which particularly stands out in the cosmological story. The reality that is the cosmos is not static, but in its unfolding displays both continuity and genuine novelty. Each stage in the story displays characteristics and properties which are emergent, that is, not reducible to the stages that preceded it. The original soup of helium and hydrogen, for example, along with the four fundamental forces of nature, is the condition for the possibility of the formation of the billions of galaxies that emerged from it; but those galaxies with their nuclear-furnace suns cannot be reduced to the stage which preceded them.

Something genuinely new and emergent had come into being. Likewise, when life emerges from inorganic compounds or when human cultures and their conscious search for meaning emerge in the story of life on earth we encounter genuinely novel realities that—while they are in some ways continuous with what preceded them—at the same time transcend them in a surprising act of creation. Creation is the emergence of new realities over against the continuity which after all makes such generativity part of a simple narrative.[13]

Everything—the stars, the earth, the myriad species and infinite forms of being, you and I, the poetry of Rilke, the paintings of Anselm Kiefer, the music of Monteverdi and of the Grateful Dead, and even this scientific understanding of the whole—has emerged from the initial Singularity. Although it is a single and interrelated reality, it is made up of a seemingly infinite diversity. As Brian Swimme and Thomas Berry describe in *The Universe Story*:

> What is particularly striking is the lack of repetition in the developing universe. The fireball that begins the universe gives way to the galactic emergence and the first generation of stars. The later generations of stars bring into being the living planets with their own sequence of epochs, each differentiating itself from the rest. Biological and human history with their ever fresh expressions of creativity continue the differentiation of time from its beginning. Indeed all fifteen billion years form an epic that must be viewed as a whole to understand its full meaning. This meaning is the extravagance of the creative outpouring, where each being is given its unique existence. At the heart of the universe is an outrageous bias for the novel, for the unfurling of surprise in prodigious dimensions throughout the vast range of existence.[14]

This is a scientific cosmology that at the same time is a religious creation story in that it leads us—as do traditional religious creation stories—to "see" or interpretively understand life "as" a meaningful whole, a unity that at the same time binds together the novelty and diversity that compose it. As Menas Kafatos and Robert Nadeau have said, "a profound spiritual awareness of unity with the whole cannot be deemed illusory from a scientific point of view."[15]

Through this cosmology nature becomes an epiphany. As we have seen, the fundamental purpose of creation mythology is to help us get in tune and harmony with a wider sacred reality, in this case the universe itself. Thomas Berry puts it this way:

> Since religious experience emerges from a sense of the awesome aspects of the natural world about us, our religious consciousness is consistently related to a cosmology telling us the story of how things came to be in the beginning, how they came to be as they are, and the role of the human in enabling the universe in its earthly manifestation to continue the mysterious course of its creative self-expression.[16]

We have in this cosmology, then, a deep and all-inclusive vision of all of creation that, like earlier creation myths, narratively reveals both a sense of what reality is all about and what the human role and destiny is within it by seeing nature as the outcome of an originating mystery which shines through it.

The religious cosmogony which accompanies the scientific cosmology is not only consistent with it but to a large extent grows out of it. Furthermore, it would seem that this cosmology precisely fulfills the nine characteristics of all creation mythology discussed in chapter one. Note first of all that it is a story. Theologian Sallie McFague tells us that

> it is a historical narrative with a beginning, middle, and presumed end, unlike the Newtonian universe, which was static and deterministic. It is not a realm belonging to a king or an artifact by an artist, but a changing, living, evolving event (with billions of smaller events making up its history). In our new cosmic story, time is irreversible, genuine novelty results

through the interplay of chance and law, and the future is open. This is an unfinished universe, a dynamic universe, still in process.[17]

This story (1) reveals a wider and deeper reality to which we belong, the whole long and vast body of the Singularity. It shows (2) precisely that all of reality is a single, meaningful and inclusive whole; and it (3) manifests the intrinsic worth of this wider reality as ultimate and that upon which everything is dependent. Furthermore, (4) it shows in precise detail how all of nature is dependently derived from and within the one in a multilevelled set of emergent modes of the Singularity. It certainly thereby (5) divulges our human rootedness and connectedness to the larger life to which we belong. As we shall see in more detail in the next chapter, (6) it stirs feelings of reverence and wonder as well as (7) a sense of gratitude that we are part of the cosmic drama. (8) It surely teaches us to be more humble in the face of such an immense reality. Finally (9), it transforms the lives of those who are touched by the story by inducing them to live in the light of the ultimate reality it narratively makes available to them.

The new cosmology, then, is both a scientific account of the evolution of the universe from its beginnings in the Big Bang and a religious creation story. It is time now to ask: What does this new cosmology tell us about the nature of God?

A child's world is fresh and new and beautiful, full of wonder and excitement. It is our misfortune that for most of us that clear-eyed vision, that true instinct for what is beautiful and awe-inspiring, is dimmed and even lost before we reach adulthood.... I should ask that...each child in the world [develop] a sense of wonder so indestructible that it would last throughout life, as an unfailing antidote against the boredom and disenchantment of later years.

§ Rachel Carson, A Sense of Wonder

God is no more an archivist unfolding an infinite sequence he had designed once forever. He continues the labour of creation throughout time.

§ Ilya Prigogine, Order Out of Chaos

We are, however, personally in agreement...that a belief in ontology, or in the existence of a Being that is not and cannot be the sum of beings, will be a vital aspect of the global revolution in thought that now seems to be a prerequisite for the survival of our species.

§ Menas Kafatos and Robert Nadeau,
The Conscious Universe

3

Wonder and the Miracle of Being

Thomas Berry, Jay McDaniel, Sallie McFague, John Haught, and a number of others have argued that this new cosmology is indeed both scientific and religious. It is a widely accepted narrative, a scientific understanding of the whole that also provides a meta-scientific, religious awareness of a reality beyond and yet inclusive of us. Cosmology affords a glimpse of that wider reality by showing that nature is the outcome of an originating mystery or what Vaclav Havel calls "the miracle of Being" which shines through it. It is this mysterious force that we directly encounter and perceive in the experience of wonder and awe that the cosmology induces in scientist and nonscientist alike. It is this sense of wonder, then, that we must explore if we are to clarify precisely the sense of divine reality which the cosmology makes available.

Existential Phenomenology

Unlike much contemporary philosophical theology, which begins with the abstract metaphysical notion of God's existence and perfection and then attempts to generate his other attributes consistent with that perfection, existential phenomenology attempts to understand the nature of God from the bottom up, so to speak, rather than the top down.[1] Existential phenomenology is the reflective description of those aspects of our experience in which we find the sacred directly present to us—in this case in the experience of wonder. The object is to articulate and make explicit what is merely implicit to the experience(s) involved. As philosopher Paul Ricoeur has put it, "phenomenology wagers that the lived can be understood and said."[2]

Such phenomenology is a first-order description, then, rather than a second-order explanation. That is, it provides a description of the experience from the point of view of the subject rather than an abstraction out of that immediate experience in such second-order hypothetical explanations as metaphysical, physiological, social, or historical "causal" accounts. Scholar of religion Wayne Proudfoot has written, "to describe the experience of a mystic by reference only to alpha waves, altered heart rate, and changes in bodily temperature is," from the descriptive point of view, "to mis-describe it."[3]

A Personal Aside

In order to put some flesh on such a phenomenological endeavor and in order to introduce the reader to the sort of experience of the holy which we shall soon be exploring, I'd like to describe a personal experience. I do this, not out of some mistaken and exaggerated sense of the significance of my particular experiences, but in order to introduce the experience of wonder and, hopefully, to throw light on the mystical experience in general which will lie at the center of our later reflections on the meaning of the new cosmology.

It was summertime, and I was fourteen years old. I had been brought up a Unitarian and—primarily because I thought of God as an entity (a man?) up in the sky and faith as no more than the superstition of fools—I considered myself an atheist. In fact, a favorite recreation at the time was attacking the arguments people used to demonstrate that there had to be such a divine entity.

Looking back on it now, it was a time of naive positivism in which it was thought science would someday understand everything and in which any spiritual dimension and understanding within human life seemed on the face of it bizarre. Religious discourse, it was often said, was simply "nonsense" since it could be neither verified nor falsified by some objective and shared experiment. Talk of spiritual understanding or truth, never mind God, was as meaningless and useless as a rush of wind in your ear or the muffled roar of the surf retreating down the rocky shore. In short, the thought of a god up and out there beyond the earth seemed about as possible to me as the idea of a giant but invisible muskrat up in the sky (in fact it still does). This sheer inability on my part to even conceive of the divine made the events which were to follow all the more surprising and significant for me.

I had had a number of accidents and illnesses since I was eight years old. Some of those illnesses nearly killed me and involved long stretches of hospitalization. In fact, the latest had occurred in the spring of my thirteenth year and involved the loss of one eye due to a BB-gun accident. As usually happens in such cases, I temporarily lost sight in my other eye out of sympathy with the damaged one, and I had my eyes bandaged and was kept for over a month in a darkened room. Sight in my damaged eye did not return and it had to be surgically removed. Needless to say, this brush with the fragility and mortality of life left its mark on me.

Another experience that reinforced this sense of the intrusion of death into life occurred the summer after I lost my eye. I had recently been released from the hospital and was taking my first walk along the shore of a local lake when I spied a beautiful heron fishing in the shallow water about thirty or forty feet from me. Its attention was totally focussed on the water. Like the unthinking adolescent I was, I thought I would throw a rock at it to see how it would react, never of course imagining that I would hit it. But, hit it I did—right on the head! To my horror, it immediately keeled over into the water. I rushed out into the water—shoes and all—full of shock and grief at what I had done and reached the unfortunate creature just as the light of life faded from its small, round eyes. Like a flash of lightning, death had suddenly and shockingly burst into the middle of life itself.

This experience along with the five or six years of recurring illness in which several times I came close to death induced in me a strong sense of mortality. Without that deep experience of my own inevitable death—without the sense of nonbeing in the form of death crouching at the boundaries of life—I'm not sure I would have come to the experience of the infinite worth of existence which was to follow. Awareness of death seems to heighten our appreciation of life.

At any rate, it was summer and I had just turned fourteen. My family and I spent our summers on Lake Waushakum (the lake where I had accidentally killed the heron) in Sterling Massachusetts. Sterling is north of Worcester. It remains a charming, small, typical central Massachusetts town with a short main street, a handful of stores, and a small central commons with a Civil War memorial in the middle. You take a right off Route 12 on Maple St. at the Town Hall, climb up a steep hill and follow a ridge for a mile or so till you come to the old Buttrick farm and orchard. Taking another right leads you down through the orchard to a magnificent hill at its end overlooking the lake. It was here beneath an apple tree looking out over the lake that my summer-long experience began. It was an experience that changed me and how I saw life and my place within it.

The first time I experienced this sense of the wonder of existence was late afternoon, as the sun was beginning to set over the lake. The sky was cloudless, and the lake reflected its high blue. I sat against one of the apple trees. The hay and wildflowers between the trees were high and the clover in the next field was in cheerful bloom. The sun was still warm on my arms and legs and I had to keep knocking horseflies off my arms before they inflicted their painful bites. There was a slight breeze that was cool on the back of my neck and stirred the apple boughs and grasses around me so that they seemed to whisper to one another like miserable old men, "What's he doing here? What's he doing here?"

I had been sitting there for awhile, not thinking of anything in particular, when I suddenly felt a rush of wonder at the spectacle before me. I don't mean "wonder" here in the sense of wondering about the solution to a problem or wondering about how to solve a puzzle. There was nothing problem-solving—or for that matter hypothetical or even intellectual—about it. On the contrary, it was a feeling or mood of astonishment, a sense of how strange and weird life is as I contemplated the sheer existence of it all, including me. "How remarkable, how strange and surrealistic," I thought to myself, "that I am sitting here in these clothes (what are clothes?) on this hill redolent with incredible insects, butterflies, and numerous forms of vegetation, overlooking a lake (strange—what is that?) with something called the sun conveniently warming me while something else called a breeze cools me. And all this is happening on a planet circling the sun at hardly imaginable speed, in a solar system hurtling through space, and within a galaxy of incredible immensity." Of course, I know now that reality is even more astonishing than I could have imagined at that time for it is all set in a universe of trillions of other solar systems in billions of other galaxies expanding away from each other at an incredible rate!

Hovering about the experience was an awareness of *mystery*. I didn't exactly say that to myself, but the wonder seemed grounded in the inexplicability of this reality unfolding before my eyes. I don't mean, here, the fact that science has not yet explained everything, but rather that such an explanation for the remarkable and astonishing existence of the whole universe— with its billions of galaxies and uncountable stars, with its extravagant and extraordinary diversity of entities (including me)—seems from this perspective *in principle inexplicable*. Even if (per the impossible) humans were ultimately to "explain" the existence of something rather than nothing by positing a first-cause or prime-mover god, we would still be faced with the inevitable and seemingly unavoidable question: *"Yes, but why does god exist?"*

Mystery here, then, isn't something thought or reasoned to, but part of the experience of wonder itself. That is, when you feel wonder you are in that very experience encountering or perceiving mystery. Reality in its myriad shapes and forms and its fifteen-billion-year production is simply encountered as amazingly strange and mysterious. It's as if the universe and our own lives are waves on a sea of mystery whose murky depths underlie and shine through everything that is.

My experience of wonder, that day, was at its core a noticing of reality itself, not this or that real thing, but the existence, the actuality, the is-ness I want to say, of everything. Wonder seems not so much an explanation or hypothesis about how or why things work as they do, but an experiential noticing or awareness of their *being* as opposed to nonbeing. It is not an attempt to "solve" the problem of existence, for example by hypothesizing a creator (which in any case just postpones the wonder as we have seen), so much as it is letting emerge into consciousness the reality of everything that is. Even if someday we could explain how everything works in a set of all-encompassing formulae or algorithms—in for example the TOE (theory of everything) which so fascinates contemporary physicists—we would still not have dealt with why everything just happens to *be*, including those formulae in terms of which the theory of everything claims to afford an understanding "of everything." Wonder is the mood or experience through which we pay attention to reality itself, to the being or existence of whatever happens to be.

There was about the experience something else that still stands out in my memory. There was a feeling of *inclusion* of myself (and everything else) within that encompassing reality, although only human entities seem fortunate enough to be aware of it. I, and the very experience of wonder I was undergoing, actually exist: I *am*. It's as if in wonder one recognizes for the first time that reality is wider and deeper than one's own petty and passing cares and concerns, that it's a bigger show than one has hitherto been aware. Reality itself, not the passing things that are, is one whole that contains all those things.

And because I and each of the myriad entities that actually exist are part of this wider, unfolding, and mysterious reality, there is a sense of *dependence* on it within the experience of wonder. If there were no such thing as actuality, then of course neither I nor anything else would be. I am a dependent manifestation of reality.

For me, the experience also contained a sense of *gratitude* for being, a sense of thankfulness for the good fortune that (mysteriously, inexplicably) I just happened to be, that I was participating in life and witnessing in wonder the extraordinary gift it is.

Finally, the experience contained within in it an impetus to live fully in the light of that giftedness, to *live* aware of it as much as possible. I had the sense, itself part and parcel of the wonder, that I had not been living as fully and deeply as I might have, that there was a certain dis-ease about my life, that I had been living a half-life distracted and unaware and thus unappreciative of this remarkable reality of which I was a part, that I was not the real

self I might be. I felt that my life heretofore had been too humdrum and practical, too much like my father: no nonsense and
all business! It seemed to me that what was left out in my experience was a sort of poetic or esthetic appreciation, a side of life
which I had inherited from my mother and which I needed to
tap in order to complete myself. Now, through the wonder and
sense of gratitude for life which went with it, I wanted to spiritually focus my life on this wider reality of which I was a small
but knowing part. It was like coming home and getting back to
what is fundamental and counts in life. I wanted to retouch and
rekindle that wonder every day, to wake up and keep my consciousness focussed upon it as much as possible no matter the
circumstances in which I might find myself. The experience left
me with a wish to live deeply and authentically, to live a *deliberative* life (in Thoreau's marvelous phrase) centered on this reality
rather than a sleeping or at least slumbering life in which I was
swallowed up in the hectic activities which seem to consume my
daily life. From the perspective of wonder, that sort of sleepy life
seemed limited by being so narrowly focussed on living that it
forgot (rather comically) to be aware of it. To live deliberatively
or mindfully means to withdraw from those narrow concerns
long enough to incorporate the awakened sense of wonder into
the very texture of daily life.

I spent most of that summer (and much of my life since)
revisiting that experience. That spot overlooking the lake became
a holy place for me, the space in which I first became aware of
the holiness of nature and all creation. In fact the experience was
fundamental to who I became and what I would do in life. For
me, as with some medievals, life itself is a sort of book (*liber
mundi*) to be read and reflectively verbalized and set out in words,
and the philosophical endeavor is the existential phenomenological task of exploring that life-experience and interpretively
manifesting its truth in words. As the philosopher Maurice
Merleau-Ponty wrote,

> The world is always already there before reflection
> begins as an inalienable presence; and all its [existen-

tial phenomenology's] efforts are concentrated upon re-achieving a direct and primitive contact with the world, and endowing that contact with philosophical status.[4]

I don't think it is farfetched to say that such fundamental and yet in-themselves mute events and experiences as my experience of wonder at Lake Waushakum await their truth—become "true," if you will—when human consciousness gives voice to them.

This experience of wonder, then, was a conversion experience, to use religious language, an experience that changed me and helped me to "see" life (reality) in a very different way than I had seen it before, to see it now as an epiphany which manifests the holy, "to see it with new eyes" as it were.

Wonder and the Mysterious Power of Being

I do not mean by wonder, here, what both Plato and Aristotle meant by it, namely a curiosity that initiates and motivates the intellectual endeavor to find answers to perplexing problems (although it does that too) and which is overcome once appropriate solutions are arrived at. Rather than intellectual curiosity, I want to claim that wonder is an experience of the radical and inexplicable mystery of being encountered at the boundaries of understanding.

As we have seen, contemporary science has brought us to just such a sense of mystery. Mystery, here, is not just what science does not yet know, but that which is mysterious in principle. Science has done this not only through quantum mechanics (Heisenberg's principle) and Gödel's theorem, which throw the rationalist enterprise of complete understanding into systematic doubt; but also through the growing recognition within science that it can never resolve such issues as the existence of the universe, the existence of laws of science at all, or even why just these particular physical laws actually hold. If, for example, science were to try to explain the emergence of the laws of science in terms of the very laws in question, we would surely be arguing in a circular manner. And yet we have no other laws in terms of which to understand their actual existence. In fact, even if we can explain how existence comes into being from a vacuum state, *that* it is and is explicable in such a manner remains radically mysterious.

We find ourselves at a boundary between what is knowable and what is in principle inexplicable and mysterious. But note, this mystery is not just the absence of ultimate scientific grounds for scientific knowledge, but a positive boundary and limit which we actually encounter in thinking and living. In discussing such mystery, scientist William Pollard observes that

> we are not speaking of a mystery of anything unknown at all. Rather we are speaking of the mysteriously amazing character of the known. There is a true mystery of the known and our modern knowledge in science confronts us with that mystery very strongly.[5]

Mysticism constitutes a core tradition within all the world's religions and is, above all, a positive and awesome *experience* of the mystery and miracle of being rather than hypothesis, inference, or mere belief about it. That contemporary science has not only made room for such mystery at the limits of reason but has actually brought us to that experience has profound consequences for religious life. First of all, it makes the scientific story of creation at the same time a cosmogony or creation myth. But secondly, by actually bringing us to that mystical experience of mystery, science may help us come to terms with what theologian Leonardo Boff characterizes as "the present crisis of the church and of the major religions," namely the "lack of any profound experience of God."[6]

Dictionaries define *wonder* as the experience of such mystery at the boundaries of life and our understanding of it. Wonder is a mood or state of curious attention and perplexity in which one notices an inexplicable and extraordinary mystery of various aspects of life as well as life as a whole.

Perhaps the broadest and most fundamental "object" of such wonder is the human experience of the astonishing and mysterious factuality of life, both as a whole as well as in each of its myriad parts. Sacred reality, in religious scholar Rudolph Otto's classic definition of the holy, is encountered as "mysterium tremendum et fascinans."[7] The experience of wonder involves a shocking and yet fascinating awareness of the mysteriousness of existence itself, *that there is anything at all*, the inexplicable power-to-be that is manifest in every entity and, indeed, in life as a whole. It is shocking because it is out of the ordinary and it unsettles and transforms our lives. It is fascinating because in the state of wonder we are transfixed by it. It is inexplicable because it lies at the boundary of or outside explanation. Foustka in Vaclav Havel's version of the Faust legend says that

> modern biology has known for a long time that while the laws of survival and mutations and the like explain all sorts of things, they don't begin to explain the main thing: Why does life actually exist in the first place, and above all why does it exist in that infinitely bright-colored multiplicity of its often quite self-serving manifestations, which almost seem to be here only because existence wants to demonstrate its own power though them?[8]

I believe that our religious traditions and indeed the spiritual life of all humankind emerge out of this jaw-dropping experience of wonder that strikes the human mind as it experiences the inexplicable grandeur of all of creation. Theologian John Haught indicates just this when he tells us that

> for religions, home in the deepest sense ultimately means mystery.... Religion is often understood as the trustful entry into an acknowledged realm of mystery.... The universe itself is an adventure into mystery, and our religions are simply ways of explicating the inherent character of the universe at the human level of emergence.[9]

The new cosmology induces in many just such a sense of wonder in the face of the mysterious coming-into-existence of the manifold of modes and stages of creation. Dennis Overbye in his history of the development of this cosmology expresses his wonder in this way:

> What could be closer to the flavor of myth than the notion that the universe did in fact appear, perhaps out of nothing; that the atoms in our bones and blood were formed in stars light-years away and billions of years ago; or that the even more ancient particles of which those atoms are composed are fossils of barely comprehensible energies and forces that existed during the first microsecond of creation? We are all artifacts of the universe, walking reminders of the ultimate mystery. We are walking dust, waking stardust.[10]

In contemplating both the emergence of something from the Singularity that constituted the origin of the universe as well as the continuing explosion of new forms and entities which emerged over fifteen billion years, we are amazed at the sheer factuality of it all. As biologist Loren Eisley says, "nature is one vast miracle transcending the reality of night and nothingness."[11]

Wonder, then, is our human reaction to the exuberant and astonishing power of things to be—that is, their sheer is-ness. In his memoir of Wittgenstein, Norman Malcolm reported that Wittgenstein once read a paper on ethics in which he pointed to just this sort of experience.

> [H]e said that he sometimes had a certain experience which could best be described by saying that when I have it I *wonder at the existence of the world*. And I am then inclined to use such phrases as, "How extraordinary that anything should exist."[12]

In the *Tractatus*, Wittgenstein tied this sense of wonder to the inexpressible existence of the world. "It is not *how* things are in the world that is mystical, but *that* it exists."[13]

More than a century ago, Kierkegaard emphasized that existence, or the sheer actuality and there-ness of things, is more than and different from any predicates we might apply to it. That is, there seems to be an unbridgeable gulf between actuality and theories and thoughts about it, or, as the tradition argued, between existence and entities (which we can describe) that are. New Testament scholar Marcus Borg puts this very well in a footnote in his study, *Meeting Jesus for the First Time*.

> By "is-ness," I seek to express a difficult but obvious notion: namely, that which "is" independently of the maps we create with language and systems of ordering. Chief among these creations are social maps based on culturally generated distinctions. These maps become the source of identity, creating social differentiations and social boundaries. But all of these maps are artificial constructions imposed upon what "is" and what we "are." Beneath the world we construct with language is is-ness.[14]

Physicist Menas Kafatos describes this distinction between determinate entities and their Being, which can only be apprehended mystically in wonder and awe, in the following way:

> Since this single significant whole, or Being, must be represented in the conscious content as parts, or beings, it is not and cannot be a direct object of scientific inquiry or knowledge. Thus any direct experience we have of this whole is necessarily in the background of consciousness, and must be devoid of conscious content....The evidence for the existence of the ineffable and mysterious disclosed by modern physics is as near as the dance of particles that make up our bodies, and as far as the furthest regions of the cosmos. The results of the experiments testing Bell's theorem suggest that all the parts, or any manifestation of "being" in the vast cosmos, are seamlessly interconnected in the unity of "Being." Yet quantum physics also says that

the ground of Being for all this being will never be completely subsumed by rational understanding.[15]

Astrophysicist John Wheeler—one of the important figures in the development of the new cosmology—expresses this same thought in his own colorful way:

> There is nothing deader than an equation. Imagine that we take the carpet up in this room, and lay down on the floor a big sheet of paper and rule it off in one-foot squares. Then I get down and write in one square my best set of equations for the universe, and you get down and write yours, and we get the people we respect the most to write down their equations, till we have all the squares filled. We've worked our way to the door of the room.
>
> We wave our magic wand and give the command to those equations to put on wings and fly. Not one of them will fly. Yet there is some magic in this universe of ours, so that with the birds and the flowers and the trees and the sky it flies!...If I had to produce a slogan for the search I see ahead of us, it would read like this: "That we shall first understand how simple the universe is when we realize how strange it is."[16]

Reality, then, is radically mysterious in so far as it transcends any possible theories or words about it. We stand before reality in all its forms in fascinated wonder and awe. Scholar of religion Barbara Sproul indicates that it is precisely the point of religious mythology to bring us to this sort of awareness:

> Religions stand before the fact of all this reality and think it wondrous. They find it awesome that it is, that it becomes and ceases to be in the same way as the whole. That star is dying, this tree is coming into being, that baby is being formed. Imagine that! (Here again the difference between myth as an expression of religion and cosmology as an expression of science is evident. While they often speak of the same subjects,

the focus of myths is on value and meaning; that of science is on facts. Both religion and science speak of moments of universal and particular creation; only religion declares them wondrous and sacred.)[17]

Jaime de Angulo, an anthropologist who lived with the Pit River Indians of Northern California, points out that the core of their religious attitudes and practices is just such wonder:

> The life of these Indians is nothing but a continuous religious experience. To them, the essence of religion is the "spirit of wonder," the recognition of power as a mysterious concentrated form of nonmaterial energy, of something loose in the world and contained in a more or less condensed degree by every object.[18]

Such wonder is a particular way of perceiving nature and life, with new eyes you might say. What formerly was seen to be ordinary and perhaps even humdrum is now perceived as *extraordinary*. Sallie McFague puts this beautifully in her recent book, *The Body of God*:

> Suddenly to see some aspect of creation naked, as it were, in its elemental beauty, its there-ness and suchness, stripped of all conventional categories and names and uses, is an experience of transcendence and immanence inextricably joined. This possibility is before us in each and every piece and part of creation: it is the wonder at the world that young children have and that poets and artists retain. It is to experience the ordinary as extraordinary. This is experiencing the world as God's body, the ordinariness of all bodies contained within and empowered by the divine.[19]

God, then, is not so much this sparrow, this chain of mountains or that galaxy or solar system, but the continuous eruption into being of those myriad forms, the active *that-ing* or *is-ing* of everything which emerges into consciousness in the experience of wonder. It is being itself, then, for as historian of religion

Mircea Eliade has said of the Sacred within archaic religions, it "is equivalent to a power, and, in the last analysis, to reality. The sacred is saturated with *being*," i.e., an astonishing power to be.[20] Medieval mystic Meister Eckhart says of God that he "is like nothing so much as being [Esse].... everything that God is is being."[21] We encounter this is-ness in wonder, astonished and overwhelmed at the extravagant profligacy of creation which erupts from emptiness into so many intricate and extraordinary forms. In cultural historian and theologian Thomas Berry's words:

> The term "God" refers to the ultimate mystery of things, something beyond that which we can understand adequately. It is experienced as the Great Spirit by many of the Primal Peoples of the world. The Great Spirit is the all-pervasive, mysterious power that is present and observed in the rising and the setting of the sun, in the growing of living things, in the sequence of the seasons. This mysterious power carries things through to their brilliant expression in all the forms that we observe in the world about us, in the stars at night, in the feel and experience of the wind, in the surging expanse of the oceans. Peoples generally experience an awesome, stupendous presence that cannot be expressed adequately in human words.[22]

The experience that Albert Schweitzer called "reverence for life" seems very close to this sense of wonder. It was during a trip through the African jungle, he tells us in his autobiography, that he first became conscious of it. On the third day of the trip, when at sunset he and his party were crossing a river amidst a group of agitated hippopotamuses, he was struck with the thought that every entity (including human beings) exhibits a reverence for and affirmation of life by seeking to maintain its existence. Reverence for life, he tells us, is a world and life affirmation in which we become conscious that "I am life which wills to live, in the midst of life which wills to live."[23] In my terms, what is sacred about all of nature is precisely this welling-forth of Being that we encounter in the perseverance of each and every entity that is.

That the ultimate no longer appears to be clothed in the arbitrarily derived terms of our previous understanding simply means that the mystery that evades all human understanding remains. The study of physical reality should only take us perpetually closer to that horizon of knowledge where the sum of beings is not and cannot be Being, while never being able to comprehend or explain this mystery.

§ Menas Kafatos and Robert Nadeau,
The Conscious Universe

The core of our human identity is nothing more or less than the fitful apprehension of the radically inexplicable presence, facticity and perceptible substantiality of the created. It is; we are. This is the rudimentary grammar of the unfathomable.

§ George Steiner, Real Presences

4

Mystery has energy. It pours energy into whoever seeks an answer to it.... I am talking about the general psychological health of the species, man. He needs the existence of mysteries. Not their solution.... An answer is always a form of death.

§ John Fowles, The Magus

Either life is always and in all circumstances sacred, or instrinsically of no account; it is inconceivable that it should be in some cases the one, and in some the other.

§ English journalist observing
the Sisters of Charity in Calcutta

The Transcendence
of God in Nature

As we have seen, God is experienced in astonishment and wonder precisely as the sheer *actuality or existence of things, a mysterious, creative force throughout the universe that is apprehended in and through the manifold of everything that is.* In more traditional religious terminology, he is transcendent. New Testament scholar Walter Brueggemann argues that this sense of wonder at the inexplicable given-ness of life, which he calls "the mystery of God," is exactly the Biblical sense of creation.

> Creation as understood in the Bible seeks to explain nothing. Creation faith, rather, is a doxological [praising] response to the wonder that I, we, and the world exist. It pushes the reason for one's existence out beyond oneself—[and finds] that reason in an inexplicable, inscrutable, and loving generosity that redefines all our modes of reasonableness.... That I exist is a reality that is referred outside myself to the mystery of God, to which I can only respond in gratitude and doxology [praise]."[1]

In traditional Western philosophy and theology, being is transcendent—that is, not reducible to entities or particular modes of existence that *are.* We say that the table, chairs, house, and universe "are," but their reality or being—their "are-ness" if you will—can never be identified with any one of them, for then in order to have the quality of being each of the others would also have to be a table, a chair, a house, or a universe. The tree "is" but that quality of existence cannot be reduced to what the tree happens to be. Heidegger classically asserts this in *Being and Time*: "The Being of entities 'is' not itself an entity."[2] In other words, if the meaning of being could be identified with a particular entity that is—or indeed even all the entities that are—then any other particular entity could not and would not *be.* Stated more simply, the meaning of being and whatever entities that happen to be are not equivalent. Being is transcendent in so

far as it is *indefinable* in terms of any determinate or finite predicates. Being and entities, then, must be separated, not in a spatial sense, but in terms of their essential meanings. The meaning
of "is" is transcendent and indefinable.

Interestingly enough, the same thing turns out to be traditionally true of God. Religiously, to think of God as some sort of
entity (or other kind of reality such as value, spirit, or love) is
idolatry in so far as it reduces "him" to (and replaces him with)
a finite something or other. God could not be an entity or even
the whole class of existing entities for then he would be a self-
contradiction, both infinite and finite at the same time. God, like
being, is *transcendent* and not reducible to particular predicates.

Some people seem to think that God is neither being nor
some particular entity that happens to be, but some sort of third
reality beyond either or both. But then we would ask, "does it
exist?" If such a third reality existed, it would have to be either
an entity (defined as some particular thing or some other determinate reality such as goodness, beauty, consciousness, or
power); or being itself. But if it were what we mean by "entity"
(and if not what in the world are we talking about?), then as we
just saw it would be a contradiction in terms. And if not an entity, then it would have to be either being itself or simply not be.

We are led by this logic to the view that God and being are
identical, as indeed I claimed in the preceding chapter. God is
the creative force of being which permeates the universe. St.
Thomas Aquinas was the most important and influential Western
theologian who held this same view that God is being itself
rather than some sort of special and all-powerful entity. Aquinas
argued further that such being must be "separated" from existing finite entities in the same way that a cause is "separated"
from its effects. If God-being were identical to the being dis-
played by all the entities that are, said Aquinas, it would entail
that God's nature is added to every time there is a new act of
being. This would contradict the simplicity and perfection of the
divine nature.[3] Therefore, God must be *separated* from existing
entities as the pure cause of their particular acts of being. In
other words, as being itself God is separated or different from

entities as such, and also further separated from the is-ness of each and all such entities as the infinite and pure cause of them. God lies outside or beyond existing nature. A number of difficulties are associated with Aquinas' notion that God's transcendence entails that he "exist" outside of or separate from nature.

- To separate God as being from entities that are is to separate their reality from them. But as Heidegger says, "Being is always the being of an entity."[4] We only encounter the actuality of entities with actually existing entities.
- To separate God-being from the act of being in which entities actually are is to confuse a difference in the kinds of reality they are with a difference in their spatial location. That is a conceptual confusion. It also leads to a contradiction in so far as it treats God who is infinite as a finite thing—only finite things after all are spatially located.
- Finally, in holding that God is separate from nature we are led to a picture of God's mystery and transcendence that seems to me is confused and false to our experience. I call this the Wizard of Oz model of transcendence. In this model, transcendence is pictured as a spatial separation of God from nature and human experience, like the opaque curtain in the throne room of the terrible Wizard of Oz in Emerald City that separates and hides from view the meek and humbug man from Omaha behind it who is pretending to be the wizard. Such a picture of transcendence seems to lead us to the untenable view that God, like the humbug wizard, is a sort of onto-theological entity that would be present to us if we could somehow remove the curtain blocking our view. As Toto pulls aside the curtain revealing the fake wizard, the wizard seeks to recover by telling Dorothy to "pay no attention to the man behind the curtain!" This idea of God is not only an idolatrous reduction of God to the status of a finite entity, but it also seems to assume that God could be defined and understood if only we could get behind the curtain that hides him from us.

Theologian Gordon Kaufman seems to take just this position. "Mystery," he tells us, "is fundamentally an intellectual term, not an experiential one…. [that] does not, in fact, tell us anything specific about the subject matter we are…seeking to understand." Of the various religious traditions, he goes on to say, "none has succeeded in comprehending the ultimate mystery of things."[5] Notice that he thinks "mystery" is an intellectual term rather than an experiential one. He goes on to talk of God as a sort of "X" beyond and behind mystery, which, like the curtain in *The Wizard of Oz*, hides him from us.[6] Notice also that he seems to think the religious quest is ultimately to *understand* God behind or beyond mystery, rather than (as I have argued) simply to accept that mystery as the human experience of God-being's transcendence and indefinability.

What can we conclude from all this?

1 If God is, he can only be being itself; he cannot be an entity of any sort.

2 As being, God is neither spatially separated from nor identical to existing entities. Rather, although always and only encountered with those things, he is ontologically different from them in so far as he is not reducible to any defining qualities or characteristics specific to any of them. The anonymous mystical *Book of Privy Counseling* puts it this way:

He is beingness, both to himself and to all. And in that way only is he separated from all that is created, in that he is both one and all, all things are one in him and all things have their being in him, as he is the being of all.[7]

This is what has been called "panentheism:" "the belief that the Being of God includes and penetrates the whole universe, so that every part of it exists in Him, but (as against pantheism) that His Being is more than, and is not exhausted by, the universe."[8] God is both immanent and transcendent.

On the one hand, modern physics precludes any reality "outside" creation, while on the other theology demands that

God be transcendent to that creation. I believe that this model in which God is identified with being is consistent with the need for an immanent form of transcendence.

3 Both supernatural theism (which separates God-being from nature) and pantheism (which collapses that God-being into some kind of entity or indeed the whole set of such entities) are untenable.

4 God is being itself.

God and Nature

As we saw in chapter One and as I have just argued from another point of view, the sacred is considered above all to be the *real*, the ultimate ground and foundation of all that is—an astonishing coming-into-being of every and all entities that happen to be. As such, it is that which is basically meaningful about existence, that in terms of which some human beings interpretively understand life and their role and destiny within it. It is, then, not itself finite or dependent for its existence and meaning on anything else; on the contrary, it is infinite, not finite. It is transcendent, as we saw, ineffable or beyond all specific and defining (finite) characteristics. To quote Lao-Tzu,

> The Tao that can be told
> is not the eternal Tao.
> The name that can be named
> is not the eternal Name.

> The unnameable is the eternally real.
> Naming is the origin
> of all particular things.

In Christianity this way of thinking about God was called the apophatic tradition or the via negativa, and was closely associated with mysticism. Here, although God is experientially available in nature and within ourselves (not separate from nature), he is at the same time transcendent (more than just nature). We can only begin to approach this holy and transcendent reality by first denying all finite characterizations of "him," including of course that he could in any literal sense be a "he." The only way to characterize God is metaphorically in such images as *light* in the Gospel of John or both *light and darkness* in the words of Dionysius the Areopagite.

> The Divine Dark is the inaccessible Light in which God is said to dwell. Into this dark, invisible because of its surpassing brightness and unsearchable because of the abundance of its supernatural torrents of light, all enter who are deemed worthy to know and see God: and, by the very fact of not seeing or knowing, are truly in Him who is above all sight and knowledge.[9]

The dominant strain of the high culture of Europe, however, developed an alternative picture of the nature of God and his relationship to nature. In this picture his transcendence is interpreted to mean that he is spatially located *outside* (and temporally *before*) his creation. This is an image that leads us to think of God as not available to human experience, a God whose existence is simply inferred and hypothesized to be the first cause of creation. This view came to eclipse the apophatic and mystical tradition and ultimately led to the spiritual numbing and disorientation so characteristic of our time.

Our European cultural tradition is a synthesis of Jewish biblical and Greek philosophical cultural values and attitudes toward life. Ancient Greek culture contributed the belief or attitude (hermeneutic or interpretive understanding of life) that what is fundamentally meaningful about life is rational understanding and explanation. Christianity picked up this Greek interpretation of what life is all about and wove it together in its fundamental theologies with the very different set of beliefs about life embodied in the biblical tradition. In this new worldview God was pictured as: (1) a kind of ultimate *explanation of things*; (2) known through *rational inference and argument*; and (3) thought to lie *outside* and *before* creation. The fascination with rational explanation and understanding inherited from the

Hellenistic tradition has contributed to the erosion of spiritual life in the high tradition of the west because it pushes God out of nature and thus beyond direct experience.

The cosmology of Hellenistic Rome conditioned how the early Christians thought about God and gods, and how that heavenly dimension was related to the earth and human redemption. The cosmos was pictured as a number of concentric spheres with the earth at the center. Beginning with the earth, there was thought to be a hierarchy of powers and spirits identified with the various astral spheres ranging from daemons to angels, gods, and ultimately the high God as source and inclusive of all the other powers, but lying outside the outermost sphere. Since God was thought to be eternal and not in motion (the Unmoved Mover), then he had to be placed outside the concentric spheres that were characterized precisely in terms of different kinds of motion and temporality.

The Jewish/Christian God was identified with the first-principle-god of the philosophers, which was thought to be the ultimate rational explanation for the universe. God was thought to lie outside the rotating spheres in a changeless state of eternity. To be consistent with this understanding of the cosmos and a first principle that pictured a grounding reality beyond motion of any kind, God was placed outside nature.

As the second person of the Trinity, Christ was associated with the Jewish Old Testament notion of Sophia (wisdom) and the Stoic conception of Logos (natural law, word, way). He thus was thought to be the mediating link between God and creation, eternity and time, spirit and matter, and God and man. Beginning in the Gospel of John and progressively unfolding in such second- and third-century Christian Fathers as Origen and Justin Martyr, the Hellenistic outlook on life was conjoined with the early Christian notion of the redemptive Christ to form a hybrid Christian theology in which God was: (1) thought to be the ultimate rational ground or explanation of the universe; (2) inferred from the character of creation itself to exist and be the "cause" of the universe; and (3) pictured as outside or separate from his creation.[10]

In the thirteenth century Aquinas brought all this together for Europe in a fateful way by adding to it the apparent biblical notion that creation actually had a beginning. God, then, was not only the *final cause* and *intelligent designer* of the intelligible order which pervades the cosmos, but was also the ultimate *first cause* who brought it into being at the beginning. As we have seen, this meant for Aquinas that God was pure being, and as such necessarily separate from his creation. We can infer from the existence and nature of this cosmos that God exists and what he is like. Every effect must have a cause and (Aquinas thought) there is always a similarity between a cause and its effect. There can't be an infinite regress of causes for that would signal absurdity and permit no ultimate rational explanation and ground for what is. God is therefore thought to be an intelligent designer and both first and final cause of nature. Finally, God must lie outside and before the universe, for if not he would necessarily be pantheistically and idolatrously reduced to a finite thing or the whole bunch of finite things. God then, (1) is the ultimate causal designer who (2) can be inferred to exist through the various proofs and who (3) must lie outside and before his creation. Notice that God is being marginalized experientially by being pushed outside the universe into a supernatural realm and by being made the object of argument and inference only.

This understanding of God as the ultimate first cause and designer of the universe was further developed in the Enlightenment, but with an emphasis upon design which flowed from that period's fascination with clocks and machinery. A nascent Deist such as Descartes subtracted God from the material earth, thereby separating him from nature. The sole relationship of God to nature, then, was to set its "machinery" going in its natural way and thereafter have nothing to do with it.

The eighteenth century English theologian William Paley later argued that if we came across a watch on a deserted beach we would naturally infer the proximity of an intelligent being capable of designing and constructing such an intricate bit of machinery. Analogously, he argued, nature was a vast machine that provided evidence for the existence of a causal designer who is known to exist by rational inference and who lies outside and before the machinery of nature. God, in effect, threw the switch that set the whole thing going so that physics could ultimately account for how it works.

Finally, Newton pictured God as the causal precondition or medium in which natural motion occurs, namely absolute space and absolute time.

> Lord God, pantokrator, or Universal Ruler.... He endures forever, and is everywhere present; and by existing always, and everywhere constitutes duration and space.[11]

Regardless of his "presence" here in a special sense, God is still pictured as the causal designer of nature who precedes it and thus lies outside and beyond it and who can be inferred to exist from the very nature of nature itself.

Direct religious experience of the divine is made impossible in favor of a view of God as an ultimate explanation or hypothesis transcendent to or separate from the universe. Insofar as he is before and outside creation, removed from nature, God is no longer available within nature and can only be inferred to exist.

In the popular tradition of Christianity, as well as in most other religious traditions, however, the divine is always available and is directly and mystically experienced in and through nature. Paul says in his letter to the Ephesians that there is "one God and Father of all, who is above all, and through all, and in you all." (Ephesians 4:6) Luther in his 1527 treatise, "This is My Body," echoes Paul, declaring that God is "in every single creature in its innermost and outermost being, on all sides, through and through, below and above, before and behind."[12]

But European tradition pushed God off before and beyond the universe and made him an intellectual explanation—a conclusion of an argument rather than an immediate, nonintellectual and nonhypothetical experience. For many people (including scientific cosmologists like Stephen Hawking) this remains the model of God and his relationship to nature which shapes the very way they talk about him. Just listen for example to the

young British writer, Robert McCrum, who describes his ago-
nizing recovery experience after suffering a massive stroke that
paralyzed one side of his body.

> I did not pray. Several visitors later asked if, having
> "looked into the abyss," I had experienced any reli-
> gious emotions—to which I can honestly reply that I
> did not. What I did find was that the world seemed
> almost unbearably precious.[13]

Strange, isn't it? From my point of view, he is describing the
religious experience in his overwhelming sense of the "preciousness" of life; but because he apparently has an entirely different conception of religion and God, he is simply unaware of
it. Strange indeed, but not surprising given the picture of God
and religion so prevalent in our culture. In fact, it is this picture
that most people have in mind when they either assert or deny
the existence of a deity "before" the Singularity which initiated
the universe some twelve to fifteen billion years ago.

Notice that all of these conceptions of the divine—from the
Ptolemaic to the Newtonian—are not only consistent with the
"scientific" understandings of the universe prevalent in their
time, but are to a large degree derived from them. Robert
Jastrow discusses this in *God and the Astronomers*.

> I think that in any age, one needs to be able to imag-
> ine the Creation in some way related to the images and
> ideas of one's own time. The writer of Genesis lived in
> a place where clay was ubiquitous; he describes God
> making man out of clay. He uses images of serpents,
> orchards, and swords which were familiar to his time.
> It is difficult for people now to see the essential under-
> lying truths through these pictures from another peri-
> od; it is an immense help to hear these truths
> reframed in the ideas of our own times, and with
> images related to the galaxies, the photons, and the
> electrons which one hears about in the newspapers or
> uses in a television set.[14]

Within contemporary science, of course, all of these views that locate the divine either outside creation or, if you will, "beneath" it as the absolute condition for its possibility, become untenable. Since space and time after Einstein are considered to be essential aspects of creation rather than conditions for it (which is to say there is no absolute time and space in which the created cosmos might reside), Newton's view simply becomes untenable. Likewise, because space is an aspect of that creation, there can be no "outside" wherein the divine can be presumed to exist. What then can we say? To be consistent with contemporary science, we must find the divine within nature yet avoid reducing it to a finite thing or even the whole of such finitude. We need a sense of transcendence that is immanent in nature but not identical to it, what American theologian Sallie McFague has called "cosmocentric transcendence."[15] As we have seen, this is called panentheism.[16]

I have argued that God must be immanent in nature, not as a finite part of it, but rather as the infinite and mysterious power-to-be that shines through it. God is the "beyond" of existence, the mysterious indefinability of reality, and it is available in nature itself. Such a God is not an explanatory hypothesis which we infer must exist, but is directly available in the experience of wonder. As we have seen, it may be that the new cosmology makes such an experience possible once again.

God, then, is the creator who "causes" the universe, not in the sense of a push cause, but by holding it in existence at every instant—that is, by being the actual existence of the universe. He is both the original and continuous creator in the sense that, in astrophysicist John Wheeler's words, he makes the whole thing "fly," or rather that amazingly enough it actually is flying! You'll recall that he argued that understanding the universe in its entirety can never make it be or "fly." Holy being isn't an explanatory hypothesis to account for the first cause of everything as much as the wondrous actuality of things, including of course the reality of the immense, fifteen-billion-year-old universe as a whole. God here is the sustainer and continuous creator of the universe, the astonishing ground or power of being from which everything has emerged. Scientist and theologian Arthur Peacocke writes:

> What the scientific perspective of the world inex-
> orably impresses upon us is a dynamic picture of the
> world of entities and structures involved in continu-
> ous and incessant change and in process without ceas-
> ing. As we have seen, new modes of existence come
> into being, and old ones often pass away. In the world
> new entities, structures and processes appear in the
> course of time, so that God's action as Creator is both
> past and present: it is continuous. Any notion of God
> as Creator must now take into account, more than ever
> before in the history of theology, that God is continu-
> ously creating, that God is semper Creator. [17]

Expressing the same view metaphorically, we could say with Sallie McFague that nature is "the body of God" and its emergent and still emerging creative order is "the mind of God." Paul Davies in The Mind of God uses just that metaphor, switching the focus from prior causality and a designer spatially external to creation to an immanent cause and designer which is at the same time transcendent or more than any and all of the entities which inhabit creation. [18]

In spite of his own intentions, the reflections of Hawking point in this same direction. Quantum theory, he tells us, has made us aware that space/time has no boundary, no edge at which we might need to fill in the lack of explanation with the primary causation of God. There is no need, then, to posit an original cause (God) to account for creation.

> One could say: "The boundary condition of the uni-
> verse is that it has no boundary." The universe would
> be completely self-contained and not affected by any-
> thing outside itself. It would neither be created nor
> destroyed. It would just BE." [19]

Hawking obviously thinks he is doing away with the notion of God precisely because he thinks of God in the traditional terms of primary causation before and outside creation. I have been arguing on the contrary that what we mean by God is precisely what Hawking asserts about Creation: it simply *is*. In fact, his position is rather close to mine, or mine to his to express it more modestly. He says, "Why does the universe go to all the bother of existing? What is it that breathes fire into the equations and makes a universe for them to describe?"[20] Notice that Hawking is still searching for some sort of rational explanation (and feels uncomfortable without it); I am simply asserting a nonexplanatory *experience* outside the dimension of explanation altogether. As philosopher Renee Weber says, "Science works to explain the mystery of being, mysticism to experience it."[21] What Hawking fails to recognize is that the sheer BEING or facticity of nature is not intelligible or explicable: it is a mystery met in wonder at the boundary of rational discourse.

I am claiming that the sacred is in fact the "object" of wonder, existence itself, and as such is perceived to be the holy, eternal, transcendent "creator" of all that is. The proper and perfectly appropriate description of this holy reality, then, is that it is that mysterious aspect of the universe which is beyond explanation and is experienced as such. It is the beyond in our midst. Rather than an entity that stands as the first cause of everything, God is the strange and unsettling *more-ness* of the existence of entities and the universe as a whole. Reality is more than you can say about it. Dostoevsky, in *A Raw Youth*, expresses this transcendent spiritual reality available in and through nature itself through the character Makar Ivanovich Dolguruby.

> What is the mystery?... Everything is a mystery, dear;
> in all is God's mystery. In every tree, in every blade of
> grass, that same mystery lies hid. Whether the tiny
> bird of the air is singing, or the stars in all their mul-
> titudes shine at night in heaven, the mystery is one,
> ever the same.[22]

Life itself is a mystery, Dostoevsky declares through this character, and that is its spiritual meaning. God is the mystery of being. To live is a wondrous thing, and so is that death that is accepted after a life of honoring God for the gift of being.

In its fascination with the sort of explanatory understanding utilized to control reality for human ends, our modern tradition has overlooked reality as such, which has always been available to the different kind of understanding involved in the sense of wonder. Modernity transformed religion into a form of explanation and God into an ultimate hypothesis, numbing us to a sense of wonder and the wider reality available through it. The religious understanding manifest in the new cosmology is closer to esthetic sensibility than scientific hypothesis. It is less a matter of explanation and control than an appreciation of life and a sense of worth and meaning in living it. See how simply and directly the Zen poet, Basho, expresses his fascination with and appreciation for life in its mundane everydayness.

> From among the peach-trees
> Blooming everywhere,
> The first cherry blossoms.[23]

Then Moses said to God, "If I come to the people of Israel and say to them, 'The God of your fathers has sent me to you,' and they ask me, 'What is his name?' what shall I say to them?" God said to Moses, "I AM WHO I AM." And he said, "Say this to the people of Israel, 'I AM has sent me to you.'"

§ Exodus 3:13-15

Is God dead, as Nietzsche claimed? Are those of us who think it worthwhile to discuss how we can think of God as out of touch with reality as the old saint whom Nietzsche's Zarathustra met in some secluded spot and about whom he felt constrained to ask: "Can it indeed be possible? This old saint in his forest has not yet heard that God is dead!" Or does it mean that we have to think anew and more deeply on what this word "God" signifies and that we have to be prepared for just as revolutionary a development in our conception of God as took place when the old mythological ideas of God were discredited and then superseded by subtler conceptions.

§ John Macquarrie, Principles of Christian Theology

5

But Does God Exist?

We now have such a long tradition of treating God as a hypothetical inference rather than a directly encountered phenomenon ubiquitous to human experience that readers are bound to ask the familiar question, "Does God exist?" What is it that we encounter in wonder? If wonder is the experience of God as the inexplicable and therefore mysterious power of things to exist, how does that happen? Obviously, we don't "see," "hear," or "touch" God in the ordinary sense of those words. What then? How do we perceive him? What are we encountering in wonder? And when it does happen, are we experiencing a reality or a mere figment of human imagination, a construct of the human need and desire for meaning projected on a mute and meaningless nature?

These questions presuppose several basic beliefs and attitudes toward life—in other words hermeneutical assumptions—that we have inherited from the Enlightenment, for the most part unquestioned assumptions that deeply characterize our modern culture.[1] A Hermeneutic, you may recall, is a way of interpretively understanding reality by "seeing" it "as" meaningful in one way or another. Needless to say, it is a far different form of understanding than the matter-of-fact understanding so important in science.

First, Descartes and the Enlightenment in general simply drained nature of any divine presence, rendering it a mere machinelike stuff. God was projected right out of nature (and experience), and thus—as we have seen—conceived of as a sort of ultimate hypothetical first cause *apart* from the universe that could only be *inferred* to exist from the nature of his creation.

Second, having drained it of any possible epiphany, nature was thought to be a meaningless stuff, a manifold of object-things that in themselves have no value or meaning. From this point of view, that we experientially encounter nature as laden with such values as beauty or the divine can only be explained as a human construct projected upon nature. If nature is simply in itself meaningless, where else could such intrinsic values

come from? "Beauty," as it is said, "is in the eye of the beholder." We now realize that this killing-off of the spirit in nature was
in reality a convenient sleight of hand. Previously, nature as a
"trace" or symbolic manifestation of God was seen as intrinsically valuable in various ways.[2] Now it was seen as a mere set of
"things" to be mined and utilized for human production and
consumption. Nature became a "natural resource," as we say
today.

Added to those two assumptions was a third. Descartes split
reality into two different kinds: the knowing subject(s) and the
known object(s). Again, that objective reality was thought to be
stripped of all secondary qualities imposed upon it by the knowing subject. The problem of modern philosophy was to show
how that objective reality "out there" got over into the mind of
the knower as a subjective representation of it. With this widespread assumption or way of interpreting life (seeing it as a
mere object devoid of any value or meaning), modern philosophy swung back and forth between a naive realism in which the
subject is simply an empty mirror that reflects objective reality
and a subjective idealism in which objective reality was thought
to be a complete construct of the knowing subject. Even its
being or reality, for Berkeley, was thought to reduce to perceptions of it ("esse est percipii").

We have here, then, three powerful interpretive assumptions: (1) God is a cause lying outside nature and thus seemingly not experientially available within it; (2) objective reality in
itself is meaningless and valueless; and (3) therefore any meaning or value that nature might be said to manifest within human
experience must be a human construct imposed on it. It seems
to me that these three uninspected assumptions underlie and
impose on modern consciousness the inevitable question of the
existence of God. "Does God exist?" seems to spring naturally to
the lips of nearly all modern people whether they be professional philosophers or students in a dormitory discussion about
life. This is in spite of the fact that the experience of God is common in most religious traditions and in spite of the fact that (as
I have already argued) God is available in and through nature as

the creative force or power to be that it manifests to human wonder.

But how does this happen? How do we perceive the presence of Being? What is it that we encounter in wonder?

Beginning with Edmund Husserl, but further developed in the writings of thinkers like Maurice Merleau-Ponty and other existential phenomenologists, the notion of an experiential reality underlying and presupposed by the Cartesian and modern separation of the thinking subject and the perceived object(s) has been deeply explored.[3] This experiential reality has variously been called "everyday existence," "Being-in-the-world," "être au monde," "carnal life," or, as I prefer, just plain "ordinary experience." "Phenomenology" is simply the philosophical method by which such ordinary and immediate experience is brought to light. We need to explore this everyday experience to discover, if we can, how the Cartesian dualistic view grew out of it and hid it from view, generating a set of uninspected assumptions that, as we have just seen, lead us to raise the question, "Does God exist?" In addition, we need to explore this reality in search of a way to get beyond that question by articulating exactly how the experience of God in wonder actually comes about. This is a sort of archeological process in which we dig through various levels of reality to find an original stratum on which all the rest are grounded and in which the experience of wonder in the face of the mystery of being is situated.

Ordinary Experience

Instead of an isolated "subject" separated from a thing-in-itself "object," we find that ordinary experience is always "intentional," always a field that includes both subject and object. There can be no knowing subject that is not knowing some sort of object (a seen mountain, a heard bell); likewise, there is no objective reality that is not an object for some subjective perceiver (a seeing, hearing, touching, remembering perceiver). In other words, the notion of subjects as well as objects in themselves apart from one another is neither experienced nor by definition possible. Descartes' belief that you can have a knowing mind in itself or that objective reality as it is in itself can be assumed is simply false to our immediate everyday experience. Objective reality is never known as it is in itself or to itself, because it isn't *to* itself. It is always *to* a subject of some sort, and vice versa. This has radical implications for our question concerning the existence of God.

Note that the manifold of ordinary objects conceived of as "the" external world is imagined and never directly perceived as such by a knowing subject. I don't mean, here, that objects don't exist. I mean simply that they are first of all perceived and then imaginatively constructed from a series of perceptions. The barn I see off in the distance is an off-brown almost grey color, but of course only to a particular perceptual eye; it certainly must look very different to a hawk with its particular biological endowments and perspective. The barn has a front and a back, but only to a perceiver who walks around it. It has a shape or form, but only of a certain kind depending on the perspective of the perceiving subject—from far away or close by, from the side, from above and so on. I can knock on the door, and it not only resists my knocking hand, but also releases a sound because of it. There is no sound as you and I experience "sound" in a world with no ears to hear it. I can actually run my tongue over the boards that compose the barn, and the taste is dry and dusty. The barn in and of itself does not *taste* "dry and dusty," nor of course could I taste the "dry and dusty" boards unless I ran my tongue over those boards. There is not and cannot be any subjective experience called "dry and dusty" without a reality external to my perception which actually tastes "dry and dusty" when I run my tongue over it. And there cannot be "dry and dusty" tasting boards out there unless and until a tongue is run over them. The dry and dusty barn boards as well as the tongue that can taste their dry and dusty character are necessary for the taste-perception of "dry and dusty."

The images I have of the barn are not experienced as mere representations over here in my mind cut off from the world they are about. Rather, in experience I see, hear, touch, and taste the barn through the barn images tied to those senses. Far from cutting us off from reality, our sense images of it precisely reveal or make available characteristics of the realities on which they are focussed.

To extend this a bit further, we human beings not only seem by our very nature to be constantly in the business of sensing/comprehending reality all about us; we also extend those senses by technological innovations that expand our abilities to explore that reality. We develop electron microscopes to try to comprehend the very minute realities within nature. And we develop space vehicles, radio telescopes, and the Hubble telescope so that we can better comprehend the larger and more distant universe. These inventions, like the bodily senses they expand and extend, neither invent realities out of whole cloth nor project upon nature wholly human understandings. Like those basic senses and experience which they expand, technological inventions permit us to discover (dis-cover or un-cover) and explore reality and thereby come to know it more deeply; but always and only from a point of view and within the limitations of human sensation and cognition.

The fact that through memory and language I come to gather the series of perceptions I have (of the barn for example) into an *imagined* objective reality (the barn) which I then can study scientifically presupposes (and then ignores) my ordinary everyday, carnal experience of it, including the fact that I can remember the series of perceptions through which I come to imagine "objective reality." What reality in and of itself is like is a mystery beyond our senses. We can never know reality in and of itself, for we have no access to such a reality. We only know it from our experiential, perceptual and conscious point of view as subjects in a world. It would seem, then, that in our ordinary experience we are not uninterested and disembodied "subjects" present to a manifold of "objects," but carnal agents always involved with and using a meaningful reality all about us. *At base we humans are entities who are always experientially up against, open to, and concerned about reality.* This is also true of the dream realities I dream and the conceptual ones I think. Dreams exist, not as the ordinary realities I perceive but as realities that are "there" for my dreaming!

The Cartesian dichotomy presupposes this real experience in order to construct an imagined "objective world" of object-things apart from a "knowing subject." This allowed the kind of empirical and deductive knowledge necessary for mastering and controlling the world to develop. Separating experience into these two kinds of reality (subjective mind and objective reality) froze these categories into separate and uninspected presuppositions in terms of which we interpret or "see" life. In effect, half truths abstracted from our ordinary experience in which a subjective perception and a perceived object are always interrelated were transformed into types or modes of reality which then cover up and hide that *actual relational experience* from our own consciousness. Here is the fatal step in which reality (being) and the world is transformed into a collection of objects present to disembodied subjects. Mind and matter become the key components of what after all is actually an interpretation of nature and life.

And that set of assumptions leads to the question of God's existence and the relationship of matter, meaning, and value.

By objectifying nature or reality and insisting that it can only be known scientifically, particularly when the idea of subjective mind apart from it was soon dropped in European culture's haste to develop technological and industrial control over nature, we gained a certain power and control that we then confused with our own destiny as human beings. But the modern haste to gain security by conceiving reality as objects present to us and knowable only scientifically was actually an interpretive understanding (hermeneutic) which ironically led to the eclipse of that very dimension of human interpretive understanding. In this way, modern culture lost not only God (now projected out of and before experience) but our own possible spiritual openness and growth. By objectifying the world, we block the path to spiritual freedom and responsibility. Heidegger calls this the world's "night" and describes it this way.

> The man of the age of technology, by this parting, opposes himself to the Open.... The essence of tech-

nology comes to the light of day only slowly. This day is the world's night, rearranged into merely techno-logical day. This day is the shortest day. It threatens a single endless winter. Not only does protection now withhold itself from man, but the integralness of the whole of what is remains now in darkness. The whole-some and sound withdraws. The world becomes with-out healing, unholy. Not only does the holy, as the track to the godhead, thereby remain concealed; even the track to the holy, the hale and whole, seems to be effaced. That is, unless there are still some mortals capable of seeing the threat of the unholy, as such. They would have to discern the danger that is assail-ing man. The danger consists in the threat that assaults man's nature in relation to Being itself, and not in accidental perils. This danger is the danger.[4]

In our ordinary experience we very rarely perceive mean-ingless objects (which of course from the Cartesian point of view is what we inevitably perceive *before* we project meaning upon them). For the chemists who deal with them, the beakers and Bunsen burners and sinks in the chemical laboratory are certainly not useless or meaningless objects simply to be gazed at. On the contrary, their meaning lies in their use. Experientially, we always encounter and perceive things and groups of things as meaning-ful in one way or another. Mount Fuji is perceived as a holy moun-tain by the Japanese; Washington D.C. is a meaningful center of American history and the Ganges for Hindus is not just a river, but a holy river that manifests the sacred journey of the human soul. Even to interpret things as "meaningless" is, strangely enough, a human way of finding meaning—in this case in the *absence of mean-ing*. It is hardly uncommon to hear the following interpretation of life: "Given the vast depths of space, the repeated supernova deaths of billions of suns, the seemingly endless generations of human dead, and the apparent demise of everything in time (even the universe), life is surely a meaningless shrug in eternity." Such an assertion of absence of meaning is (or can be) a meaningful proposition and interpretation of life for human beings.

However, I rarely experience blank objects devoid of all meaning; neither do I perceive them as unconnected instants which I then put together into a chronological series. No. I perceive my friend walking across the street to greet me, I see him approach and put out his hand, and I hear him say, "Hi, Paul. How are you?" In other words, we perceive *process* and not just individual moments which we then must add up to make an event. The world is perceived as a meaningful process and series of actions and events.

Furthermore, we are always learning more about the reality to which we are connected by carnally exploring it. A baby reaches behind the couch to find out what is behind it; I walk around the mountain to see if there is a pasture on the sunny side; a biologist explores the genetic code of chimpanzees. We are always in process, always learning more about the reality in which we find ourselves.

On the other hand, we cannot do away with that reality's ultimate mystery because we can never know reality *as it is in itself.* "Existence is always more than you can say about it." Although we can never finish our personal and species-wide journey to understanding (as we saw in chapter 2), we do know something about that reality. Drawn by the mysteries we encounter at the boundaries of our understanding, we are continuously learning more. Yet we never reach the end of the endeavor; we are always and infinitely on the way.

"External reality" is not invented or totally constructed by the human mind—either in terms of what it is like or in terms of the fact that it is at all. That's the old Cartesian dilemma again. Rather, our perceptual and conscious experience presents us seen, heard, felt, and tasted images (or their technological equivalents) that help us to understand reality more deeply. These images do not represent reality as much as they act as vehicles through which we encounter and ultimately imagine what it is like. This is called "critical realism," and it is the view that we in fact do learn about reality, but only in ways that our senses, minds, and human longings are capable of.

It may be helpful to summarize these arguments and reflections.

1 The question, "does God exist?" has been shaped by three assumptions or beliefs that emerged in the Enlightenment and that came to broadly characterize modern culture and the technological industrial world into which it has flowered: (1) God is separated from nature; (2) nature is interpreted to be a whole lot of meaningless things; and (3) reality is split into separate modes, knowing subjects and known object-things present to them.

2 Those assumptions constituted a fundamental interpretive understanding of life and our human place within it, an interpretation or hermeneutic of the meaning of being.

3 This modern interpretation or attitude toward life envisioned being or reality to be a vast manifold or set of objects that is present to consciousness and genuinely knowable only through scientific induction or mathematical deduction. ("Positivism" as it has been called in the twentieth century.)

4 Subjective consciousness or mind started out with Descartes as a form or mode of being *separate* and *different* from objects. That subjectivity was then reduced to the status of another (although "higher") kind of object, knowable like any other object or set of objects only by science. Thus being was interpreted to mean object-things present to us.

5 This modern interpretation of being, which emerged out of the world of ordinary experience, offered a set of metaphysical abstractions that then obscured that ordinary experience to ourselves. Even though we actually live that ordinary experience always and inevitably, it became hidden from our own consciousness. Along these lines, Heidegger asks at the beginning of *Being and Time* what sort of an entity a human being is that he can think of himself as an object-thing even though he never is that to himself.[5]

6 That covering-up of our own carnal experience (including the fact that such experience inevitably involves a nonscientific form of understanding—faith or interpretive understanding) led not only to our characteristically modern obliviousness to everyday experience, but also (as Heidegger puts it) to a withdrawal of the gods and a consequent emptiness, meaningless-

ness, and "night." In other words, by leveling-down the meaning of being to objects present to us and knowable only scientifically, modern culture committed itself to the conquest of reality "at the cost of losing touch with Being."[6] The more we assert our will to control the more we enter a dark night in which the holy and creative power-to-be is pushed out of experience and our need to appreciate life is simply shunted aside. God and spiritual experience are not argued out of existence; they become invisible and we become simply indifferent to them.

The cost of conquering being in order to secure our lives has been an alienation from ourselves to such a degree that God and faith make no sense at all. They are simply written off as manifestations of unscientific ignorance and superstition, as many of my students are so quick to tell me. Not only is God a meaningless hypothesis from this perspective, but the openness and human integrity that may result from the experience of God is simply not an option or a live possibility.

7 When we seriously explore our own ordinary experience, we find that we are never apart from the external world, nor is it (for us) knowable apart from the access that our senses and technological extensions of them permit.

8 We neither totally construct the world nor simply mirror it as it is in itself. We do know the world, but only within the limits of our biological and perceptual and cognitive possibilities. We are led then to a form of critical realism. We do genuinely understand some things about the world, but we can never in principle know it entirely or as it is in itself. Mystery is an inevitable framework of human experience and knowledge.

Yes, But Does God Exist?

No. God does not "exist." God is existence itself, that aspect of reality—indeed reality itself—which we discover in the ordinary experience of the opposition and resistance of otherness. The snow crunches under my feet, my muscles ache when climbing the mountain, the beautiful sunset catches my attention, I trip over a rock or slip on the ice. I discover this reality with other people: My lover refuses to be just what I want her to be. I even can and do reflect on myself as other: "My big problem is that I lack self-esteem." These and all the rest of our actual, daily experience is the mundane way we encounter and perceive this God—as the *otherness of reality*. Novelist Iris Murdoch claims that just such a perception is the basis of art and morals.

> Love is the perception of individuals. Love is the extremely difficult realization that something other than oneself is real. Love, and so art and morals, is the discovery of reality.[7]

There is another encounter with being that shows up in our ordinary experience. Whether in the context of psychotherapy or down-to-earth conversations about real problems, difficulties, emotional reactions, hopes, or fears with family or friends, we all notice the difference between an honest exploration of a person's personal reality and a "canned" version of it. You might tell me about the terrifying grip of an addiction in which you find yourself, or a set of knee-jerk reactions you fall into in the face of your lover's anger. When it is honest we say it "comes from the heart." We not only can tell when a person is genuinely touching and communicating serious realities in his or her life (concerns like death, fear of the unknown, self-loathing, anger), but when it happens it often puts listeners back in touch with similar realities in their own lives. There seems to be a deep level of honest communication that discloses and gives voice to fundamental and *real* aspects of our ordinary experience. Such honesty displays that reality for us and others to see.

Although such real communication expresses something "true" about that hitherto mute (and hence to both the speaker and listeners "unknown") and lived experience, it does not and cannot exhaustively articulate it. Part of the very experience of honest expression of our lives is that there is always more to it than you can say about it. Thus, for ourselves (and others) we are always on the way, always coming to further understand ourselves, never in complete possession of self-knowledge.

Far from an unremitting skepticism, then, our ordinary experience of external reality as well as our reflective awareness of our own everyday experience is constantly and consistently up-against and laden with being. *We are reality-conscious entities par excellence, both in our prereflective experience and our reflections on that experience.* That of course is the fundamental reason why we humans are so fascinated with being, why we seek to interpretively understand it, why we are so passionately involved in the (hermeneutical) question of the meaning of being. To say that is really to say that at base humans are spiritual entities.

And of course, knowledge depends on such mundane experiences of reality. Without them we could have no knowledge or truth for there would be nothing to know. Understanding, after all, is about reality, else it would not be true or false, would not be an instance of knowing at all.

Heidegger urges us to "attend to the presencing of things rather than to the things themselves, to the event of presence rather than to the thing present."[8] In other words, we are capable not only of noticing things about us but also of recognizing *that* they are. In the experience of wonder we notice the presence or otherness of things apart from us and apart from whatever we perceive them to be. Through our perception we encounter being by recognizing that we experience a perceiving of something that is actually present and is perceived as such. That being is a kind of resistance to which we attend, a mysterious otherness that calls us to explore it in so far as it is there.

Philosopher Erazim Kohak described this experience of sheer is-ness in *The Embers and the Stars.*

> [T]he forest enfolds you in a profound peace and there is the same feeling, the sense of unity and the fullness of life. It is not the experience of the darkened forest, the boulders, the path, or the solitary walker. All that has receded and a different reality has moved into its place, that of the fullness of Being. In such moments you sense it is always there just beneath the surface, insistent individuality of subjects and objects,

ready to rise up when the clamor subsides. You must not impose yourself upon it. But if you are willing to listen, it is there, the fullness and unity of life, the presence of Being—and it is one and good.[9]

As both Hegel and Heidegger have argued, it may be that the condition for the possibility of becoming aware of being is a prior awareness of nonbeing. For Heidegger this awareness of nonbeing comes about through our awareness of death, which leads to a concomitant awareness of the preciousness and fragility of being itself. Theologian Sallie McFague tells a personal story which relates her experience of death and nonbeing to a sense of wonder at and deep appreciation of existence itself.

> To live at all and to know it: these are the roots of wonder. I was distinctly and peculiarly human when, at age seven, I thought with terror and fascination that someday I would not "be" any longer. In that thought was contained not only consciousness of life but self-consciousness of it: it is a wonder to be alive but it is a deeper wonder to know it.[10]

It is this reality that humans perceive in wonder when they get a glimpse of it. They react in gratitude to the fact that there is anything at all, and they orient their lives about it as what is fundamentally meaningful about living. You can almost hear representatives from all our human spiritual traditions exclaim, "Boy, did I luck out! I not only exist, but I am aware of it. Whew!" So, God does not exist: Only finite things exist. The word "God" is the symbol for that which is not finite—the wondrous power of those things to actually be.

Although we cannot say existence "exists," still we have here a sacred reality that is encountered in this life as the *beyond in our midst*. In no sense is that a "proof" that God exists, of course. The mystical encounter of him in wonder makes such proofs irrelevant. This is the self-revealing and ultimate God common to the mystical perspective that flows through all of our religious traditions. Insofar as it is not an entity or thing, it is referred to variously as nothing (no-thing), empty (Sunya), neither this nor that (neti-neti), the abyss, transcendent, or non-finite (in-finite). It is more like a verb than a noun. In fact, it *is* reality, the astonishing and mysterious eruption of something (or everything) into being. What it demands of us, as film writer and producer Marty Kaplan says, is "not faith but experience, an inexhaustible wonder at the richness of this very moment."[11]

Treating God as an ultimate explanatory hypothesis, as we saw, took God right out of the universe and made him unavailable to experience. Now, thanks to modern science and cosmology, we can see the divine mystery in or through everything that is. *What could be more real than reality itself?* Can we prove that such reality exists? Maybe, in the Cartesian manner: *Something* exists, if only the belief that it does or doesn't! But is that necessary when we can actually experience it?

All right, but then we must ask (as we do of the esthetic experience) is that before which we experience wonder really just our subjective projection onto a meaningless nature—in the eye of the beholder so to speak? Is nature in its vast variety totally meaningless until narratively shaped by human beings? Would there be an awareness of the miracle of being without human awareness? Of course not, but that doesn't mean that our awareness involved in wonder creates this process of being from whole cloth.

The very question presupposes an interpretation of reality as *either* a knowing subject *or* an independent set of objects. But, it is down on the level of ordinary experience that such a question is generated. And in that experience, reality is never perceived as simply a whole bunch of meaningless objects. Entities are always experienced as disclosive of meaning and value, including the shock of their actual *being* experienced in wonder. As philosopher Karsten Harries says, nature "speaks" to us. Still, doesn't that suggest that we have projected onto nature whatever is spoken? After all, "human beings speak, we are likely to insist, not things. Things are silent, simply there. Does not all talk of things speaking read our own being into these things?"[12]

The question remains: Are the *properties* creativity, mystery, and the astonishing power to be that is experienced in wonder really there? Or can they be reduced to the human *capacity* to experience them?

We can ask the same question of mathematical and scientific "laws" of nature. Surely if human beings did not exist an awareness of such laws would likewise not exist; and yet, to say that such laws are not *about* certain properties in nature of which they are making us aware would mean that we have absolutely no understanding of nature. That seems rather far-fetched.

Let us grant that such properties—whether grasped scientifically, esthetically, or religiously—are not reducible to the natural human capacity to experience them. If we accept that, then we are led to say that the physical properties manifest in scientific understanding, the beauty of a sunset esthetically perceived, or the extraordinary and mysterious power of being encountered in the religious experience of wonder are in some sense real. (Naturally, if there were no human capacity to know them they would never be "known.")

Taken together, the experience of wonder—in which what is encountered is experienced as above all *real*—and these reflections on the distinction between human capacities and real properties, lead inexorably to the conclusion that the holy power of being is actually real. Although not simply the product of perfervid human yearning, those real properties must await human consciousness or awareness to become "true." Philosopher Holmes Rolston puts it this way:

> The world is beautiful in something like the way it is mathematical: neither aesthetic experience (in the "high" sense) nor mathematical experience exists prior to the coming of humans. Mathematics and aesthetics are human constructions; they come out of the human head and are used to map the world. This is also true of theories in the natural sciences, of lines of latitude and longitude, or of contours on maps. Regression lines (averaging out trends in data and correlating variables) and centers of gravity do not exist

in nature. But these inventions succeed in helping humans to find their way around in the world because they map form, symmetry, harmony, distribution patterns, causal interrelationships, order, unity, diversity, and so on, that are discovered to be actually there.[13]

Once again we are led to a form of critical realism. Putting it crudely, God-being (as well as the beauty of the sunset) is there, but it is mute. If no human being is present, that beauty may never be experienced. But when human carnal intentional consciousness is present, it can give voice to that mute reality by becoming conscious of it perceptually, scientifically, esthetically, or religiously. The beauty of the sunset and the mysterious power of being are neither simply "there," nor simply over "here" in the eye of the beholder. Rather, they are in between: The experience requires both the reality that is there and the consciousness that alone is able to become aware of it and to express it symbolically, sometimes in the form of a cosmological story.

Conclusion

This wider reality—the plain and astonishing there-ness of things—seems to be that which is assumed and presupposed in our arts and sciences, that which all our inquiry is about, that which we strive to bring to its peculiar truth in pigment, formula, bodily gesture, word, and deed. It is that reality, then, which constitutes the perpetual background to all our reflections and conversations, a reality which seems eternally haunted by the expectation of an account which will render it into its truth. We are constantly aware of it, though usually not in a focused way but rather as the ground of everything that is, including our own activity and experience. It is the astonishing sunset, my child's eternal and worrisome fragility, my mother and father, the sapidity of wine, and me—or rather the fact that, beyond all miracle, each of these things and more is. It is what we are all always trying to get hold of (really be!), worry about losing (this soreness in my throat is not going away), and sometimes seek to escape in the nihilation of death. It seems that this power-to-be is so close to us that we cannot see it. Blinded, like Oedipus we rage round the world finding only night and nothingness everywhere. Having "torn out" our own possibility of spiritual vision by pushing God outside and beyond the universe, not surprisingly we have lost sight of that astonishing and wondrous epiphany in nature which is always and eternally right before our eyes. Strange, isn't it? Perhaps with the help of science and its new understanding of the origins and evolution of the universe we can once again come to see.

A person will worship something—have no doubt about that. We may think our tribute is paid in secret in the dark recesses of our hearts—but it will out. That which dominates our imaginations and our thoughts will determine our lives and our character. Therefore, it behooves us to be careful what we worship, for what we are worshipping we are becoming.

§ Ralph Waldo Emerson, On Nature

6

The recognition of intrinsic value means, at the very least, that when we use other creatures for our benefit, we do so with humility, respect, and thanksgiving for these other lives.... It might [also] mean, however, that we would look at nature with new eyes, not as something to be misused or even just used, but as our kin, that of which we are a part, with each creature seen as valuable in itself and to God.

§ Sallie McFague, The Body of God

I have just three things to teach: simplicity, patience, compassion. These three are your greatest treasures. Simple in actions and in thoughts, you return to the source of being. Patient with both friends and enemies, you accord with the way things are. Compassionate toward yourself, you reconcile all beings in the world.

§ Lao Tzu, Tao Te Ching

The Word of God, our Lord Jesus Christ
Who of his boundless love
became what we are
to make us what even he himself is.

§ Irenaeus, Against Heresies

La Vita Nuova
(Life Tranformed)

As we have seen, influenced by the upper-class ancient Greek fascination with *knowledge*, the high tradition of philosophical theology in Europe interpreted Christian faith as a kind of cognitive or hypothetical understanding.[1] Christian faith thereby underwent a tidal change in meaning from personal decision and transformation in the early Jesus movement— entering the reign of God[2]—to doctrines and dogmas *about* him.

The struggle between this Hellenistic interpretation (hermeneutic) of life and faith and the early Jewish Jesus group's interpretation was fought out in the developing cultural and religious history of Europe. In the medieval period, the issue was joined around the relationship of faith and reason. The so-called Voluntarists, for example, made the will essential to the life of faith, whereas the Thomists and Dominicans in general stressed the role of *reason*. Opposed to the rationalism of Descartes, Pascal considered faith to be a matter of the heart, a way of living in the spirit of love (caritas). Still later, Kierkegaard vigorously attacked Hegel's rationalism from the point of view of Christian faith lying outside and beyond it. Indeed, for Kierkegaard living a life of reason was a form of despair and sin—an attempt to die while still living, a distraction and avoidance of the transformed life of grace. Nietzsche in turn blasted the entire western intellectual tradition as nothing but a morally bankrupt, disguised form of "the will to power."

The issue here is a hermeneutical one. That is, it's a question of what we are living for, what we find to be the wider meaning of life in which we interpretively understand our own development and destiny. The hermeneutic committed to rational understanding naturally "sees" everything in the light of that: "God" becomes an ultimate hypothesis to explain creation; "myth" is interpreted willy-nilly to be falsehoods or illusions; and "faith" is reduced to a primitive form of understanding. On the other hand, the hermeneutic committed to God involves a way of living centered on God or "the will" of God. As we saw in chapter 1, mythology is a narrative that discloses the connection of our lives to what is ultimately real (on which everything else depends) and sacred, thereby making an interpretive under-standing available. And, finally, faith in this context is not a way of knowing, but a way of *being* in the light of an interpretive understanding made available in myth, especially creation myth. From this point of view, the life committed to reason is itself a form of faith. The consuming commitment to reason is not itself rationally grounded or demonstrable (as a friend of mine puts it, "it's a no-brainer"), but rather is an idolatrous distraction from faith in God.

Furthermore, religions are not only interpretive under-standings of life that spell out a way of being, but practical pro-cedures for *transforming* life in the light of them. Mythology entails ritual and practice (and vice versa). Religious rituals are scripts for carrying out the vision of how to live contained in the origin myth underlying it. Religion has to do with practical ways of actually living. To experience wonder in the face of the amaz-ing proliferation of life all around us is to be changed—deep-ened and strengthened—to face our everyday celebrations as well as adversities. Rachel Carson put it best for me in her beau-tiful little essay titled *The Sense of Wonder*.

> What is the value of preserving and strengthening this sense of awe and wonder, this recognition of some-thing beyond the boundaries of human existence? Is the exploration of the natural world just a pleasant way to pass the golden hours of childhood or is there something deeper?
>
> I am sure there is something much deeper, some-thing lasting and significant. Those who dwell, as sci-entists or laymen, among the beauties and mysteries of the earth are never alone or weary of life. Whatever the vexations or concerns of their personal lives, their thoughts can find paths that lead to inner contentment and to renewed excitement in living. Those who con-template the beauty of the earth find reserves of strength that will endure as long as life lasts.[3]

In this chapter, I want to explore the side of religious prac-

tice that brings about transformational change in both our attitude toward life and our behavior. Religious myth—including of
course the new scientific story of creation—has practical significance of just this kind. Thus the hermeneutical vision of life
made available in the new cosmology may lead to a transformed
way of being. In Catholic thinker Thomas Berry's phrase, we may
be witnessing the end of the Cenozoic (the present geological
era) cultural attitude and sets of behavior and the emergence of
a new ecozoic (environmentally sustaining) perspective that has
radical implications for our everyday lives and behavior.

Exploring the Dimensions of Transformation

This process of transforming life is represented in various religious traditions by means of concrete metaphors derived from
our everyday lives. So, for example, the prevalent metaphors for
Buddhists seeking enlightenment are crossing a river to reach
the far shore, and climbing a mountain in order to achieve spiritual elevation. The image of spiritual life as a journey is pan-
demic to religious life. Perhaps because of their conception of
history as the arena in which the covenant and redemption are
lived out, we find in Judaism and Christianity metaphors such as
"the promised land" or "God's kingdom of heaven" at the end
of history. Life itself is pictured as a spiritual passage and transformation.

I do not mean to suggest that the various religious traditions share a single vision of the steps involved in such transformation or indeed of the actual character of the transformed life.
Human cultures and spiritual traditions are the result of long
historical interpretations (interpretations of and over against
interpretations, and so on). Our cultural traditions narratively
shape history and life in different ways and thus provide rather
different models and roadmaps concerning spiritual passage and
transformation. Various religions emphasize different images of
the sacred and different ways to live in the light of it.

Despite their differences, however, it seems fair to say that
for all these traditions the transformation involves turning away

from what is perceived to be a present, inadequate life to focus on what is ultimately *real*, turning away from self-centered concerns to what we might call, with philosopher John Hick, reality-centeredness.[4] In Judaism escape from the present state of evil and limitation to a God-centered new age of justice and love involves fulfilling the covenant by living centered on God through his eternal law. For Muslims, such a transformation involves submission or total self-surrender to Allah. For Christians, such a radical transformation entails giving up the self-centered life in favor of a complete God-centered one. In St. Paul's words, "It is not I who lives, but Christ who lives in me" (Galatians 2:20). "Therefore, if anyone is in Christ, he is a new creation; the old has passed away, behold the new has come" (2nd. Corinthians 5:17). And in both Hinduism and Buddhism, liberation demands giving up an illusory life of attachment to an everyday self and self-concerns in favor of experiencing anatta (nonself) or (which is to say much the same thing) the experience of the "real" self (Atman) as a mode of Brahman.

The steps involved in such spiritual transformation are paradoxically *both* chronological *and* simultaneous. In other words, later stages in the transformational process depend to some degree on the accomplishment of earlier ones; and yet we are never entirely finished working on each and every one of them. And it goes without saying that each step or stage of religious transformation as well as the whole set or model of such steps is conditioned and shaped by the historical, geographical, economic, cultural, gender, class and religious contexts in which they occur. Yet, given all that, it seems to me that we can talk about some seven aspects or levels involved in human spiritual transformation in general.

#1 Awareness of Nonbeing. The glory of human life is that we are aware of the miracle and beauty of our actual, day-to-day experience of living. What that means is that we are not only aware of the world about us, but also reflectively aware of that awareness. We perceive the world and (in what Husserl called "the miracle of miracles") we also apperceive and can become reflectively conscious of it. It is that awareness which seeks expression in our arts, humanities, and sciences through the human ability to make it symbolically available through language, bodily gesture and movement, paint on canvas, architecture, music, and more. But the cost of that glorious human appreciation of life is the painful awareness of our own mortality, that life is tragically and inevitably haunted by death. This is to say that before we can be aware of the beauty of life we need to be aware also of its other side—the unavoidable pain, failure, suffering, and inevitable nonbeing which permeates it. Indeed, the so-called near-death experience—however it comes about—most certainly induces in many a strong reaction in the form of wonder and appreciation for concrete and everyday existence and a sort of personal commitment not to heedlessly overlook it in the future.

As both Hegel and Heidegger knew in their different ways, neither the experience of being nor of nonbeing is possible without an awareness of its opposite. Thus the awareness of nonbeing constitutes an intitial step on the path to wonder and the appreciation of being. Putting it another way, the sense of your own mortality makes you care about living your life as fully as possible before it's gone and moves you toward deliberate spiritual transformation.

All religions—without exception as far as I can tell—include suffering and death as part of the interpretive and narrative framework of meaning which lies at their core.

Buddhism, Zoroastrianism, and Christianity—just to take three typical examples—make human pain and suffering (sometimes symbolized as "evil") part of the very core of their different spiritual teachings or "ways." For Buddhism suffering is central to human life, but can be overcome by becoming aware of its cause (attachment to passing, finite aspects of life—including oneself—which distract us from the Buddha nature) and by entering on the eight-fold path toward transformation and enlightenment. In Christianity—especially in the Pauline version—we can attain redemption from suffering and death and can achieve eternal life by participating in the death and resurrection of Christ, thereby becoming children of god by dying to ourselves and permitting Christ to live within us. Finally, in the Zoroastrian apocalyptic worldview, suffering and evil are to be overcome in both a daily and final cosmic struggle between the forces of good and the forces of evil.

Death and nonbeing are forcefully confronted in the various world religions. Nonbeing is not ultimate, but rather something to be overcome and transformed in the human journey to the primordial and holy ground of being. Sometimes this is envisaged as an eschatological kingdom of heaven on earth in which the faithful will be physically resurrected in an actual earthly paradise mirroring the original paradise from which we have fallen. Sometimes the transformation brought about by death is pictured as a spiritual immortality in heaven reserved for the faithful. Sometimes the two are paradoxically combined, as in much of Christianity. Alternately, the notion of transmigration and rebirth is a nearly universal notion in the religious views and practices of the Indian subcontinent. Even when death is pictured as simply a return to the originating power of all life, as it often is, the fact is that the spiritual life of humankind in its various forms entails both a recognition of death or nonbeing as a fundamental part of life as well as a transformative way or set of disciplines to integrate it into our lives in a positive sense.

An important matter flows from this characteristically human awareness of both nonbeing and being, of death and reality or life. Our human nature and thus the various symbol-cultures it spawns (and within which we live and raise our children) seems to comprise two different and sometimes apparently mutually exclusive aspects or tendencies.

On the one hand, we have a practical side oriented toward practical success and security in the face of our awareness of the pain and threat of nonbeing. In our daily lives we struggle and strive to survive, whether by hunting game, irrigating and planting crops, making a banking system work, establishing law-making institutions, or developing technological understandings that can contribute toward such practical goals and some small measure of safety and comfort in our lives. This side of our natures has been termed "Yang" by Taoists, "animus" by Jung, "analytic concepts" by Henri Bergson, the "masculine" by certain feminists, the "Apollonian" by Nietzsche, and more recently simply "the head." Regardless of what we call it, it is widely recognized in human history and culture as a fundamental and universal aspect of human life. In other words, it is an ontological condition of human life which cannot be transcended or otherwise done away with.

On the other hand, the awareness of the miraculousness, fragility, and thus utter worth of being leads to a very different but equally necessary and important force in our lives. It leads to a need and desire to notice, appreciate, not let-slip-away in insensitive ignorance, to focus on and care about actual being whether in the form of the world around us or our own experience as we live it through. This has been called "Yin" by the Chinese, "anima" by Jung, "intuition" by Bergson, the "feminine" by feminists, the "Dionysian" by Nietzsche, and "the heart" in today's more informal language. It is a longing to pay attention, to worship and reverence the miracle of life, to let go of the practical drive for a toe-hold of security and control long enough to experience that miracle in the state of mystical wonder. As the *Tao Te Ching* puts it,

> Know the strength of man,
> But keep a woman's care![5]

Both aspects are vital to human life for both men and women and for the cultures they create and inhabit. An imbalance in the direction of either the head or the heart leads to a dissatisfactory and unfulfilled life.

Where the balance is tipped toward the head and its need for order and security, we find spiritual disorientation and a despairing sense of meaninglessness, a sense of life half-lived. Many social and cultural commentators today (including Vaclav Havel) believe that our European cultures have tipped the balance too far toward the head and security at the cost of our hearts' yearning for spiritual appreciation and care for life.

On the other hand, tipping the balance too far toward

openness to and appreciation for real life to the exclusion of the practical can lead to a sort of childish passivity, insecurity, and outright inability to survive.

The real spiritual task for individuals as well as the cultures they live in becomes the need to *establish a balance* of these two forces. Such an integration and balancing of these two aspects is in fact a fundamental goal of the spiritual transformation we are beginning to explore here.

#2 Religious Longing. One of the things that strikes us about religious people is that many seem to have a spiritual need, a yearning for something beyond the ordinary. This is first experienced as a sort of defect or lack—as if something is missing from ordinary life. They feel that the way they have been living is unsatisfactory and incomplete, a sort of dis-ease in living, a discrepancy between how they are living and how they might or ought to be living. Accompanying that sense of incompleteness is a yearning or longing for a more fulfilled and meaningful way of being. The bridge across is a spiritual transformation in which different needs in their lives are integrated into a new life and way of being.

I have a friend who was not brought up in any particular religious tradition and would not normally talk about god or religion. These terms, in fact, are rather foreign to her outlook on life. And yet, in her middle age, she developed a sort of restless yearning, a sense that the way she was living wasn't satisfying deep down in her psyche. She felt increasingly empty, she said, and looked for a means to begin to live more fully and deeply. She yearned to be, to live fully and meaningfully, especially because of her awareness of her own mortality and the seeming "once-ness" of life. To satisfy her spiritual yearning, she finally took up Zen Buddhist meditation; and that has helped her, she says, to change how she lives, to become a more focussed and spiritually mindful person, to find a balance and integration in her life.

When you are hungry food will satisfy that need. If you have a yearning and need for friendship, then of course a friend or friends is what will satisfy that need. What is it that satisfies a spiritual need such as that of my friend?

What most spiritually restless people are seeking is, as my friend put it, a deeper and more meaningful life, a way of being which is different from and beyond the unsatisfying existence they seem to be living. They seek spiritual transformation, they strive to refocus and actually live differently.

#3 Interpretive Vision. But to attain such a spiritual transformation demands understanding interpretively what life is all about. Those who spiritually yearn seek to understand a sacred reality in terms of which they envision that transformed life. That reality is perceived by them to be wider and deeper and more meaningful than their present self-centered interests and passing desires. Spiritual longing for transformation brings up the question of what I and we are supposed to be doing in life, the question of what life means, the question of reality. Even the young child who wrote the following poem already has just such an overview of the encompassing reality of God and how we ought to live in the light of his ultimate meaningfulness.

> Goodnight God
> I hope that you are having
> a good time being the world.
> I like the world very much.
> I'm glad you made the plants
> and trees survive with the
> rain and summers.
> When summer is nearly over
> the leaves begin to fall.
> I hope you have a good time
> being the world.
> I like how God feels around
> everyone in the world.
> God, I am very happy that
> I live on you.
> Your arms clasp around the world.

I like you and your friends.
Every time I open my eyes
I see the gleaming sun.
I like the animals—the deer,
and us creatures of the world,
the mammals.
I love my dear friends.[6]

To live more meaningfully, we must have some idea of what life as a whole, human life, and my life in particular are all about. Religious people seek to fulfill their yearning for a meaningful life in terms of a deeper and more encompassing reality than their everyday wishes and wants seem to provide. In short, the urge to live fully demands a vision of what constitutes the good life and a map of how to get there.

It is this interpretive understanding that is—in the following steps—gradually brought into being. By means of this process, it comes to form the character of an individual who pursues it or the cultural outlook and values of a human group in which it becomes embedded.

As we saw earlier, it is through metaphorical images and narrative mythology—especially creation myths—that such interpretive understandings of a wider order of sacred reality to which we belong are made available. I have argued that the new scientific cosmology—besides being scientific—is also just such a creation story which links us to the ultimate reality of which we are a manifestation and part.

The Zoroastrian (and contemporary Parsi) interpretive understanding of life expressed in a variety of myths, stories, and rituals is that life is a struggle between good and evil. Life then is to be lived by joining sides with the good to overcome evil. Such a hermeneutic interpretation of life typically aims toward a particular way of being in the light of it.

In order to live more fully, then, religious people seek an understanding of the sacred or, as I prefer, ultimate reality so that they can know what they are living for. They need a vision of the good life.

The religious life develops in and through the quest for a liberating worldview and vision of the ideal. It matures with formation of a faith in a unified vision of those ethical and spiritual values that promise to guide the individual and society to well-being and fulfillment.[7]

For Augustine human nature is precisely a sort of striving and yearning for peace, happiness, and fulfillment. To achieve that is to grasp the true object of this restless yearning and thus to attain fulfillment. Of course for Augustine that true object is God. In our ordinary lives, we all too often replace that true object with pseudo sacred objects (idols) such as a material goods, wealth, or power and prestige. As he wrote in The Confessions,

> great art thou, O Lord, and greatly to be praised...for thou has formed us for thyself, and we are restless till we find rest in thee.[8]

The word "religion" indicates something like this in that it comes from the Latin "religio/religere" which means to be bound or tied to. Religious human beings are tied to the sacred in order to live life as fully and meaningfully as possible. Indeed, one scholar of religion defines being religious as a "drawing near to and coming into right relationship with ultimate reality."[9]

We develop interpretive understandings of life because without them we would not know how to live. The question then is how such interpretive understandings enter into and shape how we actually do live.

First, they provide images of what the fulfilled life is like and the motivation and attraction to seek them. Religion scholar Frederick Streng writes that

> religion is the means of ultimate transformation in two senses. First, it embodies the power for transforming human awareness; it is not merely wishful thinking, nor is it just desire. Religious truth and action are intrinsically connected with the very source of existence, however that may be defined.[10]

In their myths and rituals, religions provide requisite visions of what a transformed life connected to ultimate reality might be like.

But second, the interpretive understandings that emerge in myth and become embedded in those various religious traditions lead to specific disciplines and ritual practices to achieve the ideal life they envisage. In other words, they bring about specific steps and procedures to transform the lives of those involved. In fact, it is a broadly held assumption in religious studies that mythology always implies ritual and ceremonial practice and that such practices are in effect ways of acting out the myth on which they are founded. Thus, in the Lakota purification ritual of the Sweat Lodge (the Inipi rite) the men of the tribe smoke the sacred pipe (given to them by the gods) while steam rises from water poured on hot rocks. The steam as well as the light which penetrates the darkness at the end of the ceremony when the entrance is opened represents a purification and ritual return to and reenactment of the creation origins of all of life.[11] Such a ritual repeats and reenacts a Lakota myth of origins in much the same manner that the Passover meal reenacts the Passover origins of Judaism.

#4 Spiritual Commitment. It is not enough in the spiritual journey of life merely to correctly "understand" what the journey is all about. It if were, then a university course (or reading this book for that matter) *about* the nature of religious salvation would be an entirely adequate means to achieve such salvation. "Salvation" by whatever name and within whatever religious tradition obviously demands much more. It demands repentance and change, actual transformation of oneself and how one lives (being "born again"). Putting it another way, spiritual transformation involves the whole person: one's mind, body, social relations, will, and actual behavior and conduct in the world. To deepen your initial understanding and vision of life, then, is to change who you are and how you actually live your life.

Such a process involves serious commitment to the vision of life available in the earlier stage. This is perhaps one of the most difficult stages in the spiritual journey, for it is the point of *actively seeking* serious change and transformation. It entails a positive commitment to and increasing focus on the ideal and meaningful way of being revealed in the interpretive vision. To the degree that is achieved, it also requires a withdrawal from a life centered on finite and proximate ends such as wealth, power, recognition, and pleasure. This does not mean that one can no longer have any wealth, power, recognition or pleasure. Rather, it means that if one is to achieve an authentic spiritual life these all too often consuming, finite distractions must be set aside to clear the way for the spiritual transformation that is to follow. That's why Kierkegaard called this step in spiritual development "infinite resignation." Such a commitment is often ritually celebrated in symbolic activities of conversion and commitment. Theologian Lewis Rambo tells us:

> Hare Krishna followers, for example, shave their heads and don orange robes, thus announcing their rejection of ordinary ways of living within the United States. Conversion to Judaism requires total immersion in water before witnesses, and men must be ritually circumcised. Jewish theology affirms that these ceremonies, along with other procedures, mean that the person's previous life is abolished, and that the convert becomes a new being. Christian baptism is filled with imagery of death and rebirth. Saint Paul could write, "Don't you know that all of us who were baptized into Christ Jesus were baptized into his death? We were therefore buried with him through baptism into death in order that, just as Christ was raised from the dead through the glory of the Father, we too may live a new life" (Romans 3:3-4). Baptism, then, specifically memorializes and re-enacts that turning point.[12]

To accomplish this sort of commitment one must live in a way that is not only appropriate to the vision but also supports it. Thus, since none of us is entirely alone, we need a social context or culture and periodic ritual activities to ground the vision of life involved and to sustain our movement toward it. Indeed, without the deep immersion and social support which only a spiritual community of some sort can provide, an individual's chances of successfully attaining genuine transformation are probably rather slim.

Theologian Roger Gottlieb tells us that

> religions provide rituals—acts of prayer, meditation, collective contrition, or celebration—to awaken and reinforce a personal and communal sense of our connections to the Ultimate Truth(s). These practices aim to cultivate an impassioned clarity of vision in which the world and the self are, as Miriam Greenspan put it, "charged with the sacred."[13]

Such commitment entails living and behaving in ways that are morally and spiritually consistent with it. This means slowly transcending a life focussed on and ultimately concerned about limited and passing ends, including concern for one's own personal desires, wishes, fears, and dreams. As Havel has said, a life centered simply on one's own or even one's species' longings is "demoralizing" and inconsistent with the way of existing outlined in the spiritual interpretive vision lying at its core—especially its creation mythology.

#5 Mystical Experience and Existential Transformation.

I begin this discussion of mystical experience with a definition which we can unpack and lay out in more detail in what follows.

Mysticism is a worldwide form of religious life. Some think it constitutes the essence or core of the various religious traditions. At its heart is the so-called mystical experience which can be defined as the direct and immediate experience of ultimate reality. I have described that experience as wonder at the sheer existentiality of everything. It involves identification with that ultimate reality as manifest in oneself as well as in the rest of nature. The host of things that are, including oneself, are experienced as simply modes and manifestations of a singular, ultimate and infinite power-to-be, the dazzling fact that something and everything just happen to be. That ultimate reality has different names—the Tao, God, Allah, Brahman, the Buddha Nature, Wakan Tanka. But regardless of the name, it is directly and immediately experienced as the ultimate reality on which everything that is depends for its existence. Those things that exist are as such manifestations, effusions, or epiphanies through which we encounter that ultimate and mysterious power-to-be.

By "ultimate reality" I mean that on which everything depends for its actuality, but which itself is not dependent on anything else. As Meister Eckhard wrote, "God in things is activity, reality and power." Hindus call it SAT, reality as such. As religion scholars Denise and John Carmody have pointed out:

> If there is a predominant imagery or language in which mystics the world over have tended to work out their desire and articulate what they have experienced and want to urge upon others, it is the imagery or language of being. For the Indo-European world religions, those indebted to India and Greece, thinking of ultimacy in terms of being, is-ness, seems to come naturally. Thus Hindus, Buddhists, and many Jews, Christians, and Muslims have worked ontologically, with a strong inclination to stress (1) the relation between the source or ground of the world and the world itself and (2) the dependence of the second, limited partner on a grant of being from the first, unlimited partner.[14]

Because everything is experienced in wonder as a dependent mode and manifestation of an ultimate and singular reality, the various things that exist are interrelated or parts within a single whole, reality itself. Each and every thing is interrelated and interdependent on every other thing. This is what Buddhists call interdependent causation. The ocean and rain are interdependent: Neither can exist without the other or, for that matter, without water. The entire set as well as each member in the set is dependent on reality itself insofar as it mysteriously just happens to be.

The mystical experience is a direct and immediate awareness of being. It is not a hypothesis or any sort of knowledge "about," but a direct encounter of existence in wonder. It is more like esthetically experiencing the beauty of a sunset or Mozart's Requiem than an analysis or conceptual understanding of those experiences. And just as no analysis or explanation could ever replace the experienced beauty of the sunset or Mozart's Requiem, so too no description or explanation can ever replace the actual mystical experience of wonder in the face of reality itself. As the Zen saying has it, "the finger pointing at the moon is not the moon."

There is a self-verifying element within the mystical experience: It is experienced as truly an encounter with what is ultimately and fundamentally real, neither an illusion nor a subjective projection upon reality.

> In the mystical experience, mystics are aware of ultimate reality at first hand with such vividness and such vitality that there is no room for doubt. Afterward, when "ordinary," nonmystical consciousness has returned, the mystic may reason about the experience, may even doubt what happened, but during the experience itself, there is no doubt. Ultimate reality is experienced as indubitably present.[15]

The experience is of a reality that is "transcendent" and ineffable, not in the sense of being located temporally or spatially before or outside creation, but rather in the sense of being beyond all finite predications or understandings. It is not an entity or "it" of any kind but the astonishing, indefinable, and mysterious fact that there is anything at all. As Kierkegaard recognized, existence escapes any sort of conceptual or essential definition. God cannot be known, but can be (and is) directly experienced in wonder. There is about life an inescapable mystery, a mystery which cannot be resolved or done away with, a mysterious power that is encountered at the boundaries of science and within the mystical experience itself.

Finally, mystical experience changes who you are, what you think you are doing, and how you concretely live. In a real sense, then, the experience is the heart of spiritual life insofar as it is the event that empowers a person to actually live differently. It existentializes the vision of life and reality made available in the myth and ritual of any particular religious community.

Through this experience, mystics find an identification and rootedness within a wider and deeper reality which seems miraculously and mysteriously to be always emerging from nothing. The individual stops her self-centered willing, and lets go and lets be the flow of life (being) in an act of trust. Thus, the experience is one of opening oneself to new possibilities, an opening that is essentially a trusting and hope-filled way of being. Life is transformed by becoming grounded in ultimate reality itself.

#6 Spiritual Integration: the Transformed and Transparent Life. All of these steps and stages of spiritual transformation—but especially mystical experience—lead to what I like to call spiritual integration. William James in his *Varieties of Religious Experience* attempted to give what he called a "composite photograph" of such a spiritually transformed life.

> 1 A feeling of being in a wider life than that of this world's selfish little interests; and a conviction, not merely intellectual, but as it were sensible, of the existence of an Ideal Power...
>
> 2 A sense of the friendly continuity of the ideal power with our own life, and a willing self-surrender to its control.
>
> 3 An immense elation and freedom, as the outlines of the confining selfhood melt down.
>
> 4 A shifting of the emotional center towards loving and harmonious affections, towards "yes, yes," and away from "no, no," where the claims of the non-ego are concerned.[16]

There seem to be at least three fundamental elements involved in such spiritual transformation and integration.

First, by means of the mystical experience of identification with ultimate reality— again, the miracle of being—a person is integrated into that wider reality which pervades and is manifest throughout the whole fifty billion galaxies and over twelve to fifteen billion years. Spiritual life entails an identification with the universe as we have come to know it as an emerging, creative, and fertile unfolding through the seemingly boundless forms of reality which (to borrow Sallie McFague's metaphor in her book, *The Body of God*) constitute the body of God. The deep ecology movement seems to take just such a position.

> Most people in deep ecology have had the feeling— usually, but not always in nature—that they are connected with something greater than their ego, greater than their name, their family, their special attributes as an individual.... Without that identification, one is not easily drawn to become involved in deep ecology.... Insofar as this conversion, these deep feelings, are religious, then deep ecology has a religious component.[17]

But spiritual integration also involves an integration of that ultimate reality into one's self. As Hinduism holds, Atman is Brahman—you yourself at your spiritual core are a particular, reflectively conscious form of that unfolding reality. Or, in Christian terms, to repent is to become reborn as a child of God, to die with and rise again with Christ in you. Augustine in one of his sermons to the newly baptised ("infantes") tells them the hidden (from the uninitiated) meaning of the Eucharist. They themselves, he says, *become* the bread and wine which they receive. "If, then, you are Christ's body and members, it is your own mystery that lies here upon the table of the Lord, and it is your own mystery that you receive.... It is *what you are yourselves*."[18] In contemplating the relationship between evolution and Christianity, a contemporary theologian writes,

> the scientific pictures [of the evolution of nature], it seems to me, support the biblical view that man is an image or reflection of the cosmic reality which creat-

ed him, and man has been made one with it, partner or steward in the program of bringing about the kingdom of advancing life.[19]

You are a kind of creative becoming within the encompassing becoming that constitutes the whole of reality. In Kierkegaard's beautiful expression, "Faith is: that the self in being itself and in willing to be itself is grounded transparently in God,"[20] in the power that created you. You are the Buddha nature. This is not the everyday self of just these particular perceptions, fears, hopes and desires, but a deeper and more fundamental self that has been so touched and transformed by its identification with God that it is liberated to be a creative becoming which reflects the ultimate reality lying at the heart of the universe. As theologian David Burrell explains, this is what Karl Jung meant by "individuation." Burrell starts with Jung's words here:

> "One can explain the God-image...as a reflection of the self, or, conversely, explain the self as an imago Dei in man. Both propositions are psychologically true, since the self, which can only be perceived subjectively as a most intimate and unique thing, requires universality as a background."
>
> The point Jung wishes to make is strictly analogous with that regarding metaphysical assertions. The point of God language lies in affirming that one does not create himself, but is a part of a larger context which provides significance to his life by inviting him out of a separate individuality into full-blown individuation.[21]

One important result of this integration of holy being into our lives is that we strike a balance (or restrike it) between the two different aspects or tendencies of our lives we discussed earlier. The transformed and transparent life entails a balanced life of equilibrium between animus and anima, yang and yin, male and female—between the active attempt to achieve security and control in a parsimonious nature (to survive) and the equally strong drive to find a meaning and purpose in our lives through worship or appreciation and care for living.

Finally, spiritual integration leads to a newly integrated temporal self. Our ordinary experience is structured temporally by a series of actions which aim at achieving certain short-term and long-term meaningful goals. For example, I drive down to the grocery to shop or sit down at my computer to finish this chapter. Longer-range goals can include such things as achieving financial independence, becoming a serious painter, or (more to the point) repenting and being forgiven by God. Such long-range purposes and goals constitute a spiritual interpretive understanding or clarity concerning what life is all about. Those meaningful ends reach back through our personal experiences to structure them as a narrative plot—"me," what I care most about and what makes me the particular me that I am. Theologians Stanley Hauerwas and James McClendon call "character" what I am calling "self" and think of character as built upon and pervaded by ultimate "convictions." As McClendon writes, "to have convictions is to have at least that much character."[22] Regardless of how we name it, if I dedicate my life to gaining and keeping wealth, for example, my actions will display a character or involve an "I" who is just that person who is actively seeking to become wealthy and whose life is permeated and structured by that conviction or vision of things. When a person mystically experiences ultimate reality in the state of wonder, on the other hand, a change occurs in which an attachment to finite goals (wealth, power, prestige, rational understanding, pleasure) is replaced by an attachment to what is precisely not finite, the infinite. The self and the interpretive story which structures and constitutes it are transformed. Now, transparently grounded in the creative power-to-be from which one has emerged, a new self or character is constituted founded in that universal story. In St. Paul's symbolic terminology, such a person dies to himself and is reborn in Christ. My story comes to be seen as a part of a wider story of reality itself, from beginning to now and from end to end. My story becomes revised in the light of the larger story of the universe in such a way that it takes on and reflects the creative force, the power-to-be, that is the engine of process and growth on both levels.

When I find it meaningful to live and willfully act toward finite ends, I am living a life "as-if" those goals were genuinely ultimate and nonfinite, a life distracted from God. In Kierkegaard's terms I am living an illusion and sin, or a life of despair in which I replace God or ultimate reality with an idol. This is to avoid choice in and about life by reducing possibility in my life to just this particular illusory and distracting quest, thereby excluding other possibilities for living, including living centered on the genuinely ultimate reality that is God. Such a way of living is an attempt not to live while still living. Thus, it is for Kierkegaard a "sickness unto death."

When I am living transparently grounded in what is genuinely ultimate reality, however, I am living as a self that is not attached to a finite purpose or meaning, but is instead open to possibility itself and thus is creatively becoming. At the core of our human nature is a passion to grow. In a sense, to be grounded on what is transcendent is to be grounded in "nothing" or a reality which is "empty" and thus to be a self-determining and creative becoming. Thomas Moore writes, "Faith is a gift of Spirit that allows the soul to remain attached to its own unfolding."[23] In discovering the source of life you discover who you are, you discover your own meaningful role and destiny within the ultimate reality that constituted you and of which you are a self-conscious manifestation. This integration of ultimate reality into your everyday life and concerns, then, is a kind of liberation that is accompanied by a creative energy variously referred to as salvation or enlightenment. As Harvard theologian Gordon Kaufman notes, it involves a "letting go."

> Faith is the "letting go" (Kierkegaard) of all attachments, including specifically and especially our religious and theological attachments, because it is just these idolatries which shield us from—and thus close us off from—that ultimate mystery in which both our being and our fulfillment are grounded.[24]

This penultimate stage of spiritual integration entails living a life of real freedom and creative choice over against the backdrop of the fundamental miracle of being that pervades the universe. Once again, in Kierkegaard's words, "the more choice, the more self."[25] Or one might say with McClendon, the more one has convictions the more one has character. If you live centered on (convicted by) God as possibility, then your character or self is grounded in and pervaded by possibility. You own yourself.

Theologian and scientist Arthur Peacocke reinforces this idea of spiritual transformation when he writes,

> So the "good news" is all about living our lives in and with God; about being taken into the presence of God and being reshaped after the image of Christ so that God begins to 'take over' our inner lives and our humanity begins to become what God intended. Thus we become a God-transformed being.[26]

In discussing Albert Schweitzer's conception of "reverence for life," Steven Rockefeller clarifies the affirmation of life that constitutes such "a God-transformed being."

> [I]t is instructive to reflect on the ways in which the practice of reverence for life is a path to inner freedom and to religious faith and relationship with God. Regarding the achievement of inner freedom, the ideal of reverence for life involves a commitment to be faithful to one's inmost self, to one's heart and its ideals, and this is the beginning of release from the world—that is, from a blind attachment to external things that enslaves the human spirit. On the path of reverence for life, a person begins to realize that the key to fulfillment and meaning is not what I have but what I am.... The way of reverence for life involves a faith that is religious in nature. A healthy life-affirming religious faith involves a trust in the enduring meaning and value of life in spite of all the suffering and inexplicable evil that are encountered in existence. At the core of such a faith is a great Yes to life that wells

up out of the depths of our being, possessing our minds and hearts. The experience of being grasped by the mystery, beauty, and inherent value of life is an encounter with the sacred.[27]

#7 Compassionate Love and Caring. Finally, the spiritual integration which results from the experience of wonder leads to a state of love and compassion for being in all its myriad forms. It is important to recognize that, in the words of the classic Hindu text, *The Bhagavad Gita*, "The liberated person sees himself in all beings, and all beings in himself."[28] And Alice Walker in *The Color Purple* has her character Shugg put it this way:

Anyhow, he say, you know how it is. You ast yourself one question, it lead to fifteen. I start to wonder why us need love. Why us suffer. Why us black. Why us men and women. Where do children really come from. It didn't take long to realize I didn't hardly know nothing. And that if you ast yourself why you black or a man or a woman or a bush it don't mean nothing if you don't ast why you here, period...I think us here to wonder, myself. To wonder. To ast...The more I wonder, he say, the more I love.[29]

This love is not a kind of ethical code or set of rules as much as it is a *state of being* in which one cares and takes responsibility for the body of God, if you will. By identifying with the miracle of being discovered in wonder one is transformed, becoming open to the miraculous fragility of every particular manifestation of that being. One locates oneself within the encompassing purposes of God. That in turn leads to an ethics of compassion and transformation. Philosopher Ken Wilbur looks at such an ethics as explored by Emerson and Schopenhauer.

And what has that to do with morality? Everything, according to Emerson and Schopenhauer, for in seeing that all sentient beings are expressions of one Self, then all beings are treated as one's Self. And that realization—a profound fruition of the *decentering* thrust of evolution—is the only source of true *compassion*, a compassion that does not put self first (egocentric) or a particular society first (sociocentric) or humans first (anthropocentric), nor does it try merely in thought to act *as* if we are all united (worldcentric), but directly and immediately breathes the common air and beats the common blood of a Heart and Body that is one in all beings.[30]

What seems to be required to change how we behave toward other people as well as nature is something beyond the mind and merely conceptual understanding—a mindful compassion, a caring transformation of the heart, a spiritual awakening. We need an identification with the cosmos to support such transformed behavior toward all of nature. Warwick Fox puts it this way:

> ...cosmologically based identification means having a lived sense of an overall scheme of things such that one comes to feel a sense of commonality with all other entities (whether one happens to encounter them personally or not) in much the same way as, for example, leaves on the same tree would feel a sense of commonality with each and every other leaf if, say, we assumed that these leaves were all conscious and had a deep-seated realization of the fact that they all belonged to the same tree.[31]

We shall return to the particulars of this approach to this in our concluding chapter.

If theoretical reason in modern physics does eventually refashion the terms of constructing our symbolic universe to the extent that it impacts practical reason, then conceiving of a human being, as Einstein put it, as "part of the whole" is the leap of faith that would prove most critical. It is only in making this leap that we can begin, as he suggests, to free ourselves of the "optical illusions" of our present conception of self as a "part limited in space and time," and to widen "our circle of compassion to embrace all living creatures and the whole of nature in its beauty." Yet one cannot, of course, merely reason or argue oneself into an acceptance of this proposition. One must also have the capacity, in our view, for what Einstein termed "cosmic religious feeling."

§ Menas Kafatos and Robert Nadeau,
The Conscious Universe

7

Whatever evaluation we finally make of a stretch of land, no matter how profound or accurate, we will find it inadequate. The land retains an identity of its own, still deeper and more subtle than we can know. Our obligation toward it then becomes simple: to approach with an uncalculating mind, with an attitude of regard.... To intend from the beginning to preserve some of the mystery within it as a kind of wisdom to be experienced, not questioned. And to be alert for its openings, for that moment when something sacred reveals itself within the mundane, and you know the land knows you are there.

§ Barry Lopez, Arctic Dreams

Here's the Story

I t might be helpful to step back long enough to summarize the argument so far. ¶ I began by arguing that Vaclav Havel is right to call for a new relationship "to the universe and its metaphysical order" and a "respect for the miracle of being" to ground moral, political, and spiritual meaning and orientation. The widespread spiritual and moral numbing that results from the lack of such a relationship has led such thinkers as ecological theologian Thomas Berry to contend that

> a radical reassessment of the human situation is needed, especially concerning those basic values that give to life some satisfactory meaning. We need something that will supply in our times what was supplied formerly by our traditional religious story. If we are to achieve this purpose, we must begin where everything begins in human affairs—with the basic story, our narrative of how things came to be, how they came to be as they are, and how the future can be given some satisfying direction. We need a story that will educate us, a story that will heal, guide, and discipline us.[1]

¶Recent developments in our scholarly understanding of religious mythology—especially creation mythology—have led to a new appreciation of the place of interpretive understanding in the formation and maintenance of human cultures and traditions. I have claimed that creation mythology was the traditional vehicle by means of which individuals and cultures in the past discovered a wider reality to which they belonged and through which they found their meaningful role and destiny in life. I argued that the bifurcation of life into two levels (the sacred and the ordinary or profane) within the creation myth affords a "grammar of interpretive understanding" by means of which humans come to see or interpretively understand their ordinary lives in the light of that deeper, ultimate and holy reality. By means of double metaphors, they encounter the sacred reality as, for example, a fertile mother or a loving father (first metaphor), and then "see" ordinary life "as" a dependent reflection of it. Thus life is interpreted to be a fertile birth process or a matter of accepting god/father's love or commands (second metaphor). In this way they discover a basic and inclusive meaning in terms of which they orient their lives and establish cultures pervaded by it.

¶Recent developments in science (quantum physics, astrophysics, geology, and biology in particular) have given rise to an astounding scientific account of the origins and development of the universe, often referred to as "the new cosmology." This has led in the past ten or fifteen years to a growing and very interesting conversation between scientists, philosophers, and theologians concerning the significance and cultural implications of the new cosmology.

The fact that such a conversation has not existed for the past three hundred years surely makes this conversation significant in itself. The changes in our understanding of both science and religion may help to heal the rift between what C.P. Snow called the "two cultures," that is the sciences and the humanities, or—more importantly—between our heads and our hearts. Finally, the cosmology itself may lead to a cultural shift in human self understanding, labeled variously "postmodern" or "ecozoic," a shift which some observers believe will be as important as the cultural transformations constituted by the advent of the Christian outlook in fourth-century Rome or "modern" developments during the European Enlightenment.

¶We then went on to set out the scientific story of the universe that has emerged over the past forty or fifty years. That universe is a story, or at least is grasped scientifically as such a story. Thus it meets the first criterion of creation mythology.

The implications of this are important. We humans are uniquely story-telling creatures. We tell these stories in the form of novels, dramas, folktales, and history. The stories that constitute religious mythology connect us to origins and thereby to ultimate and inclusive interpretive understanding or meaning. It may be that the foundation and condition for this human phenomenon lies in the fact that each of us in our everyday lives is intimately involved with story. Experientially, *who* I am, my character and identity, is tied up with a unique story—the events of my particular past, my present situation, and the meaningful goals I hope to achieve in the future. When we reflect on our experience, we find it narratively structured with beginnings, a middle, and possible ends, actions and events strung together in a plot that we can to a degree articulate. If you want to get to know me, for example, I would have to tell you how I became a person interested in living a spiritually grounded life, the steps I took on the way to that goal, and the present state and degree of my progress toward that goal. Such meaningful goals enable us to construe all the actions and events leading up to and toward them as a single and significant whole, in much the same way that the conclusion of an ordinary plot within any sort of human story ultimately synthesizes the events leading to it into a meaningful narrative. Stories give connective meaning to otherwise isolated and in themselves meaningless events. Each of us is an unfolding story, a meaningful plot that gathers the succession of our actions into a meaningful whole: "me." We see this plot in the light of an ultimate spiritual purpose or goal we are seeking to become. If you want to know me you must not only learn about my past and present actions and events, but also about what I fundamentally care about and want to make out of my life. You need to hear the story of my life.[2]

The same is true for human cultures. It now seems that we are finding such intimate narrativity embodied in the universe to which we belong. Just like ourselves, the universe has a biography, is a story that has emerged from the past and points forward to a meaning which has not as yet come into being. The stories of our individual and cultural lives, then, are not strange and alien realities in a universe devoid of narrativity. On the contrary, the narrativity that makes us who we are seems to parallel and reflect the larger story of the cosmic reality in which we are embedded. In a remarkable way, the book of nature seems to be reflected in our own story of trying to discern where our lives are headed and what they are all about.

¶Furthermore, as we have seen, the cosmological story manifests and makes available a wider reality, the fecund and creative power that has unfolded over the past twelve to fifteen billion years.

That wider reality which the cosmology manifests to our sense of wonder is the remarkable, fascinating, and extraordinary *power-to-be* that nature in whole and parts exhibits, the sheer *is-ness* and *that-ness* of things. Commenting on a recent conference of Benedictine and Buddhist monks in Rome, Father Mayeul de Dreuille suggests just this. He describes how these two contemplative and mystical traditions intersect precisely at their shared *experience* of the infinite ground of all being.

> In common we all believe that the infinite is the source of all being.... And so, when we enter into deep silence in our heart, we can in some way reach this source, this spring of all being, and through this source be in communion with the whole universe, unite ourselves with this infinity, this great source of peace and happiness.[3]

The cosmological story reveals a wider reality to which we belong, a reality par excellence that is ultimate and against which, like earlier creation stories, we can see the significance and purpose of our own lives. This universe becomes the context for understanding our own lives. We now can see the joys and pains of all life in the light of this cosmic story, and we can connect our own sometimes disjointed lives to the divine reality that seems to shine through it from its beginnings some fifteen billion years ago until right now. This new creation story provides that "wider reality" which many believe we need to summon into our lives.

¶The story discloses that all of reality is a single, meaningful and inclusive whole from which all the different parts of the cosmos are derived. As ecologist Warwick Fox says,

> Even if our present views on cosmological evolution (including phylogenetic and ontogenetic evolution) turn out to stand in need of modification in crucial respects, we still have every reason to believe that the

particular views that supersede these views will be entirely in conformity with the far more general idea that all entities in the universe are aspects of a single unfolding reality that has become increasingly differentiated over time.[4]

¶The universe is an ongoing and unfinished event, a radiance of revelatory and shattering creative mystery. We are not only in that, we also *are* it. We are all not only the result of the original Singularity in its fifteen-billion-year unfolding, but as well we are that reality in the particular human and cultural mode we just happen to be.

¶The conscious awareness that lies at the core of our nature has permitted the universe in a certain sense to become aware of itself. Through the new scientific story of creation, we have become aware of the mysterious, creative power which nature manifests in its incredible evolution. And since at least forty thousand BCE (It's from this period that anthropologists have found human remains buried with religious care and ceremony in caves in the Alps), humans have been religiously aware of the same fascinating and mysterious creative power. We humans have emerged in the evolution of nature and are a relatively unique species within it that is scientifically and religiously aware of the reality to which we belong and from which we have evolved. May we not say, then, that nature and the God that shines through it as its creative power and process has become self-conscious?

Being reflectively conscious of our own experience of spiritual growth is another way we become aware of the miracle of being. That is, we are conscious of our own religious longing and creative striving to achieve a meaningful future. We are aware that spiritual growth is not just an unfolding of the person each of us has always been, but is a creative process or power within us which goes beyond that to a new and transformed state of being. Thus, our reflective awareness of our own striving to be parallels the creative unfolding of nature itself, but now in a self-conscious manner. Like all of nature, then, that striving is an epiphany, but this time in the form of our own creative experience of personal growth. It is the deep within the deep, the deep miracle of being manifest in nature now aware of itself in our self-conscious striving to be. It is, as it were, the miracle of being known from the "inside." Because of this, it seems to me, there have been traditionally two routes to the mystical experience of God—an external way and an internal way.

We have a basic thirst to understand origins, to find our place in the whole, to explore reality and our own journey within it. One's story is connected to the larger story of the universe as a conscious version of it. Through the central creative dynamic of the universe, myriad forms and modes of reality have evolved. This of course includes human beings. And as far as we can tell, we humans have long been spiritually and culturally interested in the origins of life and our particular relationship to them—thus the central place of creation mythologies within various religious traditions. In the twentieth century, that interest in origins seems to be met in scientific cosmology that narratively sets out the long evolution of the universe from the Singularity till today. We study the heavens so that we can understand those origins and our destiny within it. The new cosmology, like traditional religious creation stories, helps us to see and (in a peculiar mystical way) "understand" our own being, the being of other aspects of nature, and, indeed, the ultimate being of the whole universe. The question for us in the wake of our new scientific understanding of origins is how we might best utilize our creative potential in the light of the reality which that cosmology makes available, to develop a human story that makes sense within the larger, creative story of the universe.

This story incorporates human life into the fifteen-billion-year sequence of creative transformations in our universe. Science, now, is not an enemy of the human spirit, but an expression of it. Science helps us to join the ancient human enterprise of connecting ourselves to the ultimate, fundamental and inclusive reality and living in the light of it. Human cultures around the world are beginning to tell this story. By locating themselves within it, those cultures are beginning to discover a wider reality.

¶This story also manifests the *worthiness* of that wider reality insofar as it is experienced as fundamental and ultimate. It shows precisely how all of nature is dependently derived from the *one* and it divulges our *rootedness* and *connectedness* to the larger life to which we belong. It surely stirs feelings of reverence and awe by inducing a sense of *wonder* in the face of what Havel calls "the miracle of being." It stimulates a sense of *gratitude* in the face of the whole adventure and, lastly, it may provide the foundation for a postmodern or ecozoic culture.

Accompanying the wondrous perception of the immense and awesome encompassing reality which is the universe is a sense of belonging to a wider and deeper reality beyond our shifting, practical daily concerns and a sense of gratitude that we not only are part of this immense drama but, thank God, *aware* of it!

¶Simple *intellectual* understanding of the nature and evolution of the universe is not the point here. The point is to *feel* that universe, *experience* it in wonder and awe. In other words, the goal of this new creation story—like traditional creation stories—is to bring those who understand it into an inner state of transformation. As we have seen, the ultimate purpose of religious practice is to change how we live, to transform us. We noted in the last chapter the seven steps or stages involved in this process of spiritual transformation. That transformation (termed "metanoia" in Christianity) culminates in an identification with the fecund power-to-be that is manifest throughout nature and results in a state of compassion or love for each and every entity that has emerged from it over the vast expanse of creation.

The function of creation stories in traditional religious cultures is not only to make available an interpretive understanding of a wider reality and our role within it, but also to launch us into active behaviors and practices that can transform us in such a way that we come to live and behave that vision of holy life. In other words they help us to find our way to transformed and transparent lives of love and compassion for the mysterious reality of which we are part. So too, I have argued, insofar as the new scientific cosmology is also a religious cosmogony, and insofar as it is effective, it may change us. Science writer Timothy Ferris asks:

> Who are we, and what do we want? Cosmology like every other human endeavor comes back to us in the end, but it's not about us. That's the beauty of it—that we return from the voyage altered. Galaxies, like ocean coral, work a sea change, and make of us something rich and strange...to find our place, we must know the place, cellar to ceiling, from the taproots to the stars, the whole shebang.[5]

Having a map, so to speak, of where we are in life is a first step in the process of inner spiritual transformation in which we come to live centered on the wider reality revealed in the creation story.

¶It is worth emphasizing at this point that the new cosmology does not constitute a new religion. It is not offered as an entirely new and novel religious vision of life and set of ritual practices aimed at transforming our lives outside and beyond the traditional religions. New forms of ritual worship and behavior appropriate to our time and place will undoubtedly develop. But that has happened throughout the history of our traditional religions. Rather, this new story of the universe and our place within it is meant to renew the spiritual insights and practices lying at the core of those traditions. It will probably not replace those earlier creation myths and rituals of reverence and worship as much as provide a scientifically valid, narrative lens through which we can interpret those earlier stories. Thus the new cosmology may help us to retouch, renew, and relive the spiritual wonder and wisdom they once embodied and from which we have become so alienated.

It may be that the easy separation between science and religion which I described in the introduction is in process of being overcome. Perhaps science and religion—far from being primordial enemies—have always been engaged in the same project: To understand the reality in which we find ourselves in order to live as fully and deeply as is possible. Science now can be thought of as revelations in which nature is permitted to show itself to human understanding in new ways that deepen and increase our sense of wonder and awe. Far from being deadly opponents, science and religion are increasingly becoming allies: Science becomes a way to make the sacred mystery available to religious experience; and the religious sense of wonder at the incredible mystery of being provides the motivation and drive for the further pursuit of science.

Through quantum physics, biology, and astronomy, scientists are discovering what is ultimately real—the Singularity—which has unfolded mysteriously into the trillions of unpredictable forms which constitute the universe; and they have set out in the past decades to inform us about it.

¶Finally, I have tried to show how the new scientific cosmology shows us a universe in which we have an origin, an inclusive home, and a meaningful destiny. That cosmology, I have claimed, is also a religious creation story which displays God as the numinous quality available through the story to the human experience of wonder and awe—the sheer existential fact that it is and that we are. To scientifically understand how the galaxies formed, or how life evolved from the earliest forms of Prokaryotes, or to know that the newborn baby is what she is because of her genetic code does not prevent our reaction in wonder to all those realities. Scientific understanding does not take away our awe at life. In fact, as we have seen, it *adds* to it. To explain is not to explain away. The wonder we experience before the mysterious emergence of all of reality over such a vast stretch of time enables us not just to master and control nature, but also to find our meaningful home and destiny within it. Science and religion, it would seem, are not enemies or even opponents, but rather ways in which human beings meet their different needs to understand and to find their meaningful place in a wider reality.

It may be that the great mission of our time will be to reintegrate our lives and cultures into the wider and ultimately meaningful reality which has emerged over the past fifteen billion years. That reality is not only the origin, support, and framework of our lives, but our destiny as well—a destiny in which we give both scientific and artistic voice to that mute reality as well as spiritual reverence, love, and care.

Then Moses went up from the plains of Moab to Mount Nebo, to the top of Pisgah, which is opposite Jericho, and the Lord showed him the whole land.... The Lord said to him, "This is the land of which I swore to Abraham, to Isaac, and to Jacob, saying, 'I will give it to your descendants'; I have let you see it with your eyes, but you shall not cross over there."

§ Deuteronomy 34:1 and 4

We have left the age of miracles behind, but not, I trust, our sense of wonder.

§ Chet Raymo, Alone with a Sense of Wonder

The Meno concludes that virtue does not come by nature, nor is it teachable, but comes by divine dispensation.

§ Iris Murdoch,
Metaphysics as a Guide to Morals

There is a process of ever-widening identification and ever-narrowing alienation which widens the self. The self is as comprehensive as the totality of our identifications. Or, more succinctly: Our Self is that with which we identify. The question then reads: How do we widen identifications?...In certain forms of mysticism, there is an experience of identification with every life form, using this term in a wide sense. Within the deep ecological movement, poetical and philosophical expressions of such experiences are not uncommon.

§ Arne Naess,
Identification as a Source of Deep Ecological Attitudes

8

Within Sight
of the Promised Land

We are living on an obscure but beautiful planet in an equally obscure solar system on the Orion arm of the Milky Way. There are billions of other suns in this single galaxy, not to mention the universe. The earth of course is our immediate home, and it seems to have provided an appropriate milieu for our evolution and the recent opportunity to understand the more encompassing universe of which we are part. Like Moses and his people wandering in the desert, we seem unquestionably on the way, on the way toward something, perhaps a fresher and brighter future, perhaps our own form of "promised land."

Think of poor Moses! He was not permitted to enter the Promised Land toward which he had led his people over the decades, but could only see it from atop Mount Nebo in Moab, east of the Jordan. Successfully entering into that land of milk and honey was to be left to Joshua and the younger generation that had matured in the hard desert since the flight from Egypt. Crossing the Jordan was the culmination of many years of wandering with what must have seemed the slim hope of God's covenantal promise to hold on to. Poor Moses. He (and his generation we might surmise) had risked so much, had dared to flee the cruel but in some ways comfortable bondage of the past in Egypt for the merely hoped-for future. Now he had to let it all go. He had worked so long for this moment, and yet was only given a glimpse of all that he and his people had struggled and longed for over those many years.

Like Moses, our generation may not be permitted to cross over into that land full of promise toward which we have striven so long. We may be fated to see it only from afar, as it were. We have struggled to free ourselves intellectually and spiritually from the limitations and bondage bequeathed to us from our past; and we have at least caught a glimpse of the promising cultural landscape of the future. The developments of modern science have been astounding and significant beyond measure. Likewise, our understanding of the parameters of spiritual life has certainly grown. Both have contributed to a recognition of the human promise in connecting—through the contemporary, scientific story of creation laid out in the new cosmology—to a wider reality in which we may rediscover our place and home in the scheme of things.

It's time now to draw some conclusions that flow from our explorations so far. In other words, it's time to lay out some of the promises of the promised land that our preceding reflections point to.

Spiritual Significance

We saw in the introduction that Vaclav Havel (and many others) believe that our task today is to discover a spiritual sense of a "wider reality" or "metaphysical order" to which we belong. The tragedy of modern life is that we have repressed and suppressed any idea or experience of a reality transcendent to ourselves that might provide a higher and more inclusive meaning for our lives. We have replaced that reality with our own authority, authority that distracts us from our spiritual possibilities. We then use this illusory commitment to ourselves to decry the pain and meaninglessness we ourselves have brought on and envisage political liberation and human rights to be the singular goal of all of creation. Remarkable!

To perceive the all-embracing "miracle of being" is to go beyond the constricting, nihilistic, and demoralized confines of our modern, technological, consumer societies and to find a meaningful human role and destiny within nature as a whole. Traditional religions have always been involved in this human issue of interpretively understanding how and where we fit in life. Furthermore, creation myths in those traditional societies were a fundamental means of narratively connecting the ultimate sacred ground of being with the rest of nature, including human beings. The contemporary scientific story of the evolution of the universe precisely parallels those earlier cosmogonic myths and is thus both scientifically and religiously significant. The new cosmology makes available the wondrous, inexplicable, and miraculous power-to-be of everything that is. Encountering such an ultimate spiritual reality leads to a transformation in real individual and cultural life and behavior, and to an openness and compassionate concern for all of creation. What spiritual lessons can we draw from this?

First, we should note the obvious. There is an ultimate reality of which we are part and which we encounter in the experience of wonder. It is a reality that is consistent with contemporary science insofar as it is neither "outside" nor "before" creation, but an overflowing power-to-be that is manifest within it. This mysterious reality provides a sense of purpose and meaning in life, precisely what Havel seems to mean by a "metaphysical order" or "miracle of being" beyond our individual or merely human whims and desires. Physicists Menas Kafatos and Robert Nadeau draw just such a conclusion from their work on quantum theory.

> [A] belief in ontology, or in the existence of a Being that is not and cannot be the sum of beings, will be a vital aspect of the global revolution in thought that now seems to be a prerequisite for the survival of our own species.[1]

Ecological theologian Thomas Berry makes the same point in terms of his ecological vision of "the viable human."

> Ecologists recognize that reducing the planet to a resource base for consumer use in an industrial society is already a spiritual and psychic degradation. Our main experience of the divine, the world of the sacred, has been diminished as money and utility values have taken precedence over spiritual, aesthetic, emotional, and religious values in our attitude toward the natural world. Any recovery of the natural world will require not only extensive financial funding but a conversion experience deep in the psychic structure of the human.... Both education and religion need to ground themselves within the story of the universe as we now understand this story through empirical knowledge. Within this functional cosmology we can

overcome our alienation and begin the renewal of life on a sustainable basis. This story is a numinous revelatory story that could evoke the vision and the energy required to bring not only ourselves but the entire planet into a new order of magnificence.[2]

So, our lives need not be meaningless and disoriented ones, focussed on short-term, myopic and self-centered needs and longings. Far from being absurd, the questing spiritual life of faith is not only consistent with scientific understanding but in fact seems to be supported by it in so far as that science points to a remarkable power-to-be of all things in creation, including ourselves. This entails finding our human role and destiny within that unfolding universe. As television documentary producer Angela Tilby has said,

> [R]ather than seeing the universe as the background for the human adventure, we should see ourselves as part of its adventure. The whole cosmos is the adventure and our human journey is part of that, rather than it being merely part of ours.[3]

This raises an important question. What is the correct relationship of humans and thus human culture to the rest of nature? Perhaps there is no single answer to this question, but surely one of the broadest and deepest responses resulting from the new understanding of the unfolding of the universe is that human beings are *the reality of nature now conscious of itself*. We humans find in nature a wider reality to which we belong and in that identification we can creatively discover our relation to it. Finding where we come from in nature may help us to discover our place and destiny within it. Furthermore, since the evolution of the universe is an expansion and *ascent to more complex integration*, can we not say that it provides a real basis for optimism? How can we despair when we become aware that reality—including ourselves—creatively evolves and ascends? But, mind severed from body, culture from nature, head from heart is a spiritual and ecological catastrophe in the process of happening.

As we have seen, we are *stories* within the larger story of reality itself. We are not only modes and manifestations of reality; we are also conscious of that reality both in nature and in our own experience. We are the depth and consciousness of being within encompassing being. God is available within and without.

The process of transforming our lives so that we can live centered on God means to live openly, to let go of our consuming distractions in order to let the mysterious power-to-be emerge in our personal growth. Living a spiritual life means living "transparently," to use Kierkegaard's beautiful phrase. To focus life on the ultimate sacred and miraculous actuality of things is to let go, to not live attached and reduced to a fundamental concern for this or that finite something or other. It is to live focussed on the infinite, the transcendent and ineffable coming-into-being that is reality. It is, then, as Kierkegaard realized, to become oneself, to be a process open to possibility. In the words of both Paul and Augustine, faith is nothing but living with hope and love. Let me repeat that: *Faith is a way of living with hope and love!* In Thoreau's careful language, living authentically entails living a "deliberative" life which one chooses and owns rather than being drawn here and there willy-nilly by one's immediate and finite longings and goals. Our own unfolding lives mirror the unfolding power-to-be at the heart of the universe itself.

Seeing ourselves as part of a wider reality rather than the other way around has profound spiritual and cultural implications. Theologian Sallie McFague puts it this way.

> Once the scales have fallen from our eyes, once we have seen and believed that reality is put together in such a fashion that we are profoundly united to and interdependent with all other beings, everything is changed. We see the world differently, not anthropocentrically, not in a utilitarian way, not in terms of dualistic hierarchies, not in parochial terms. We have a sense of belonging to the earth, of having a place in it, and of loving it more than we ever thought possible.[4]

Finally, it seems to me that this spiritual vision can be helpful in the dialogue between religious traditions in an increasingly small, well-armed, and interdependent world. For those traditions are, from this point of view, simply different narrative and symbolic ways that human cultures have developed to communicate this interpretive understanding of the ground of being and to foster ways to live in the light of it.

The New Story in Education

At one time, the traditional Jewish and Christian story of how the universe came to be and how we fit into it provided both a factual account of the origin and nature of the universe (as understood at the time) and an interpretive understanding of our place within it. That account of creation, of course, is no longer scientifically acceptable in any literal sense.

The demise of that conception of the origin and nature of the cosmos left our culture with a scientifically ungrounded, interpretive creation myth. That in turn led to a dysfunctional culture in which scientific knowledge and spiritual understanding were disconnected and presumed to be in opposition to one another. "Myth" in this context came to mean factually untrue or false. A gulf opened up in our culture and selves between scientific fact and spiritual meaning and value, between our heads and our hearts. In effect we were given a choice between living in a scientific and technologically powerful universe that is pointless; or living in a warm, fuzzy, and meaningful world as revealed in the creation mythology of Genesis that is after all scientifically false. Naturally, this painful separation in us shows up in our educational institutions and practices. How could it fail to do so?

It shows up in the limited modern story that informs our education at every step. In this story we make ourselves the central value and point of all creation while conceiving nature to be a mere backdrop or stage for human redemption and liberation. From this narrow perspective, human history is envisaged to be the progressive domination and control of nature for the sake of our own security and ease. Human knowledge is limited to scientific and mathematical forms of understanding, while art, literature, and religion are thought to be "cultured"—that is important—but nonetheless merely subjective or emotional projections upon a meaningless nature.

In this modern context, education is dedicated to helping our young learn the skills and understanding requisite for successfully living in such a technical world. The natural sciences and various technical disciplines like engineering, business, law, journalism, and political science are themselves broken into separate disciplines that are dedicated to such professional ends. In the meantime, there is no inclusive story that encompasses both nature and human culture, thus ultimately fragmenting education into separate and self-absorbed technical fields and disciplines. If there is an implicit story lurking about in this modern context it is that of the progressive human conquest of a silent but now less brutal (because more controlled) nature. Until recently, however, that story was more implicit than explicit insofar as the separate disciplines were not tied together into a meaningful narrative whole. The heft and focus in this educational context is upon the "real" knowledge that accrues from those scientific and technical disciplines.

We need a new story to heal the breach between scientific understanding and spiritual life—between factual understanding and normative judgments of meaning and value—so that our young may live whole lives again. Fortunately, we seem to have just such a story that lays out the origins and interconnected evolution of a universe that, if my argument has been at all successful, is both scientifically and spiritually valid.

As we have seen, scientific cosmology outlines the unfolding of the universe over twelve to fifteen billion years. It is a story. As such, it is not just a chronology of events, but a historical plot that ties into one meaningful whole the disparate events from which it is constituted. It is a story about the evolving universe that contains everything that is, including human cultural awareness of it through art, the humanities, and the sciences. It is not a story of a limited mechanistic universe as the eighteenth century presumed, but a story of emergence, novelty, creativity and thus human wonder and awe. When we experience wonder at the plethora of stars splashed across the night sky or at the earth photographed from the moon, we get some sense of the deep and creative mystery that pervades the universe. The universe of which we are part is one, and everything that over time has come to be has come from it and is inextricably bound up with it. Human life, history, and culture are such novel and unique manifestations of the original "Big Bang," and can now be seen to play a meaningful role within the encompassing whole.

As was the case for creation myths in earlier traditional cultures, it would seem that the story of that mysterious and fecund unfolding that has emerged into our human consciousness over the past fifty years has profound significance for the education of our young. It is of course a story that should inform and pervade the education of our young at all stages, for it is through it that they can learn of their place within the broader reality of the universe. The new cosmology is a way for them to learn about the development of human culture and its place on earth and within the encompassing universe. It is a way to see the big picture, to see reality and our role within it as a whole. Naturally, the very laying out of this cosmic story entails learning all the forms of human understanding that have brought about this glimpse into the nature of the wider universe, from the various sciences to art, the history of religion, and literature. How this story is to be told at different levels of education is still an open question, of course, and certainly there are and will continue to be intellectual disputes concerning the details and even the meaning of the story. But that it provides a new opportunity to restructure and reform education seems indisputable.

Imagine, for example, a core curriculum that begins with considerations in physics and astronomy of the early stages of the universe, including the intellectual revolution that we call quantum theory. After that might come courses detailing with both the history and methodological manner of understanding the geological, biological and botanical developments on earth. Courses that follow might outline the early evolution of hominids and Homo Sapiens—a fascinating and remarkably robust set of contemporary disciplines. The archaeology and history of human cultures with their diverse religious perspectives and practices, economic and political institutions, and oral and literary traditions from their Neolithic beginnings to modern history and cultural developments might follow in rich detail.

What educational advantages might this new story have?

- First, it is both a new picture and the *big picture* that frames and embraces everything that is, has been, or ever will be. In earlier cultures such creation stories formed the framework and content of education, and that may come to be in this case as well.
- This story *integrates* the fragmented disciplinary bodies of knowledge into a single, inclusive and astonishing, narrative understanding of the whole.
- Since this is a story of the universe with both scientific and spiritual significance, such an education may enable all of us to come to view life itself as a *spiritually meaningful whole*.
- It is a story of the evolution of the universe that is both *scientifically and spiritually valid*, a story that brings together our heads and our hearts. Thus, it may enable our children to live whole lives again.
- This story may illumine the natural and historical situation in which we find ourselves, thereby helping us to *discover meaningful roles* for our future. We cannot fully predict the future, of course, but in a careful and informed manner we can choose the contributions we make to bring it about.
- Finally, this is a vision with encompassing and yet diverse significance for the human family. Since it is beginning to be told across diverse cultures around the world, it may help to bring into being a new global and sustainable human world.

Love and Ethics

This spiritual vision suggests a startlingly different kind of ethics from those usually discussed in contemporary philosophy. It is an ethics of love and compassion based on the ultimate and sacred power-to-be. I'd like to call it, "transformational ethics," for it is an ethics that flows from the transformed spiritual state of integration and compassion. It is an ethics of character and virtue, then, a way of being rather than a way of thinking about morality or of intellectually justifying particular moral judgments and acts. The liberated or good person in both Hinduism and Buddhism lives in a state of mindful compassion. For Christians, redemption means living in a state of love beyond rules and law. In his first epistle, John equates being saved with existing in a loving manner: "Beloved, let us love one another: for love is of God; and he who loves is born of God, and knows God. He who does not love does not know God; for God is love" (I John 4.7-8). And Paul says, "the fruit of the Spirit is love, joy, peace, long suffering, gentleness, goodness, faith, meekness, temperance: against such there is no law" (Galatians 5:22-3). Notice. There is no appeal to principle or law here to ground and justify moral decisions. Rather, morality flows from character, "love, joy, peace, long suffering," from a transformed way of existing that is the result of "the Spirit."

It may be that just as the eighteenth and nineteenth centuries saw a cultural transformation in the powers of human understanding in the Enlightenment and the Industrial Revolution, so we may now be entering a new kind of cultural transformation. But this new and emerging enlightenment—although built on the knowledge of the universe which science is providing in the twentieth century—may have more to do with spiritual and moral rather than merely epistemic transformation.

This is an ethics based on the spiritual vision I have outlined here and is based on the mysterious power-to-be perceived in wonder at the unfolding universe science has made available to us. What does this ethics look like?

Such an ethics would not be what so much of contemporary philosophical ethics seems to be. It would not consist of attempts to resolve ethical issues and justify particular acts by rationally grounding them in an abstract principle—the categorical imperative, for example, or some form of utilitarianism. It's not that intellectual analysis is not an important component of ethical decision-making, but recent ethics seems to make ethical judgment (as well as justification of such judgments) an almost exclusively individual and "intellectual" activity.

It seems to me that a sea-change in how we think about ethics is being suggested by the new cosmology. This perspective entails an ethics that moves away from individual judgments to social contexts, away from abstractly justifying acts to feeling concerns, away from intellectual judgments to transformed character, and away from a humanistic and secular perspective to a theocentric one in which we live transformed and transparently centered on the mysterious power-to-be.

Along with such feminist thinkers as Carol Gilligan and Nell Nodding, transformational ethics emphasizes that moral judgment in our everyday lives is a social process that rests on real feelings of mutual care and trust rather than intellectual arguments and justifications. In wrestling with serious moral issues, I find that I rarely decide them alone, but always along with my family, friends, lover, and peers. Even when I do make ethical judgments by myself, it is still in inner conversation with others. And I (we) don't step back and find a principle which can then make the moral decision for me. Instead, I am in a real social context with others, and our moral decision flows from that context of care and trust.

Rather than abstract and philosophical analyses and justifications, ethics suggested by the new cosmology puts personal transformation and compassionate character at the heart of morality.

Finally, such an ethics is theocentric rather than humanistic or anthropocentric in that it is based on an encounter in wonder of the wider power-to-be that pervades nature. Thus, it is an ethics which includes natural entities beyond human beings in its moral considerations. Humans, of course, are not excluded from moral consideration since we are obviously an important part of the natural economy. Since God or the power-to-be pervades all of nature, what Sallie McFague calls "the body of God," we can now think of that nature as *intrinsically valuable*. Compare that to the typical modern view which thinks of nature as a huge set of meaningless object-things whose value is merely *extrinsic*— in other words only valuable to the degree that they have utility for humans.

Hegel, in the *Philosophy of Right*, offers an ethical perspective that advances us toward our goal by avoiding both the radical individualism as well as the hyper-intellectualism of so much contemporary ethical reflection. His ethics is not based on an individual judgment of moral right and wrong reducible to a general principle. Rather, moral judgments are social in character, involve feeling, and their ultimate goal is self-actualization and freedom within the context of a specific human culture.

For Hegel, the moral subject can only attain his or her free self-actualization within the broader social context in which he or she is embedded. He argues that an individual is a self-constituting being who is shaped by the political economy and its morals in which she matures and lives. The highest stage or rational state of "individual self-actualization consists in participating in the state and recognizing it as such an end."[5] The subjective spirit finds its realization in objective spirit insofar as it finds individual and free opportunity to actualize itself within the constraining social and moral situation in which it has grown up. This does not mean that the individual is simply historically and culturally determined, but rather that subjective spirit can only find opportunity for its particular fulfillment within a concrete social and moral set of conditions.

This is fine as far as it goes, and as my colleague Ken Westphal points out, it does not mean (as is often thought) that Hegel is an out-and-out organicist (or fascist) who recommends submersion and subordination of the individual within the organic social whole of the state.[6] But still, this is far too anthropocentric to ethically fit the wider reality manifest in nature that science is now showing us. The attainment of human self-actualization and freedom in human culture does not reach far enough. It ignores the wider reality manifest in and through nature from which—science tells us—it has emerged and on which it is dependent. It is humanistic, if you will, rather than appropriately theocentric. One might argue that the wider philosophy/theology of Hegel in which absolute spirit becomes conscious of itself in human history and consciousness comes closer to fitting the bill on a deeper level.

If we could retain the positive elements of Hegel's ethics but at the same time replace participation in the human state with participation in the wider web of reality of which both the individual and the political economy are simply parts, we would have the basic groundwork for the sort of ethics which seems to flow from the new cosmology and the mysterious reality it manifests.

Interestingly, Albert Schweitzer argues in his memoir Out of My Life and Thought for just such an expansion of ethics from its limitation to the human dimension to life itself.

> The great fault of all ethics hitherto has been that they believed themselves to have to deal only with the relations of man to man. In reality, however, the question is what is his attitude to the world and all life that comes within his reach. A man is ethical only when life, as such, is sacred to him, that of plants and animals as that of his fellow-man, and when he devotes himself helpfully to all life that is in need of help. Only the universal ethics of the feeling of responsibility in an ever-widening sphere for all that lives—only that ethic can be founded in thought. The ethic of the relation of man to man is not something apart by itself: it is only a particular relation which results from the universal one.[7]

Philosopher Steven Rockefeller argues that Schweitzer himself thought that the life force manifest in nature is intrinsically valuable insofar as it manifests the sacred.

Things have intrinsic value because they are members of the great community of life and the divine mystery is at work in them. In Schweitzer's philosophy what gives things intrinsic value is the presence of life, and he writes that "life, as such, is sacred."[8]

This is a helpful step beyond Hegel, but we must note that it limits such intrinsic value to life forms, excluding for example geological formations from "reverence for life." Such contemporary "deep ecologists" as Arne Naess and George Sessions are often understood to be equally guilty of a merely biocentric sense of the intrinsically valuable. In fact, however, they expand what is intrinsically valuable from merely life forms to reality as a whole. Such an expansion is most consistent with my emphasis on the wider power-to-be manifest in and through all of nature and encountered in the mystical experience of wonder. In their famous eight-point definition of the principles of deep ecology, Naess and Sessions's first point is that

> the well-being and flourishing of human and nonhuman life on earth have value in themselves (synonyms: intrinsic value, inherent worth). These values are independent of the usefulness of the nonhuman world for human purposes.

In a commentary on the eight points, they insist that they do not mean that only "life" forms are instrinsically valuable. On the contrary.

> RE (1): This formulation refers to the biosphere, or more professionally, to the ecosphere as a whole (this is also referred to as "ecocentrism"). This includes individuals, species, populations, habitat, as well as human and nonhuman cultures. Given our current knowledge of all-pervasive intimate relationships, this implies a fundamental concern and respect.
>
> The term "life" is used here in a more comprehensive nontechnical way also to refer to what biologists classify as "nonliving": rivers (watersheds), land-

scapes, ecosystems. For supporters of deep ecology, slogans such as "let the river live" illustrate this broader usage so common in many cultures.[9]

Deep ecologists thus argue that every life form, using this term in a wide sense, is intrinsically valuable and not to be considered of mere utility (or extrinsic value) for human beings. The emphasis here is on the *experience* (as opposed to argument, belief, or theoretical justification) of intrinsic value in the wider natural reality to which we belong and with which we *identify*. In Naess's words:

> I'm not much interested in ethics or morals. I'm interested in how we experience the world.... Ethics follows from how we experience the world. If you experience the world so-and-so then you don't kill it. If you articulate your experience then it can be a philosophy or religion.[10]

For Naess and other deep ecologists human self-actualization is attained only by identification with the wider nature to which we belong. Rather than finding its fulfillment in human society, self-actualization for them is possible only within the web of nature, within the laws, rules, and actual interrelated behaviors of all aspects of nature in its active unfolding.

This introduces a problem. If everything is equally intrinsically valuable precisely because it *is*, then how can we make moral choices between them? Does this mean that the Ebola virus is just as holy and intrinsically valuable as, for example, human beings who might contract the disease and die from it? Does this mean that we should not take sides in this struggle by seeking means to defeat the viral infection in humans or even to eradicate it entirely? Would it mean that we ought not to avoid grizzly bears or vaccinate our children against tuberculosis because it would prevent inherently valuable entities from achieving their full potential?

I think it is true that anything that exists *is* intrinsically valuable for just that reason, including grizzly bears and the Ebola virus. That means that when we must choose between two forms of being, as choose we must, then we should do so with compassion and (in Kierkegaard's phrase) with some "fear and trembling." If we must choose between the life of a particular person or the existence of the Ebola virus about to attack him, then of course we will probably choose to save the person (unless, perhaps, it was Hitler). We are forced to make such moral judgments between contending goods and as living entities we have a built-in biological will to live. But against the backdrop of identification with and concern for being in all of its manifestations, we should do so hesitantly, concernfully, and with some anguish at the step we need to take.

It seems to me we face a question of not just individual entities in conflict, but entire species as well as biotic regions and contexts in which they proliferate. Given our identification with being, we would probably hesitate all the more to destroy an entire species or biotic region sustaining many such species. Although each individual entity is intrinsically valuable as a manifestation of God, an entire species seems all the more intrinsically valuable. Yes, I believe there are degrees, or shall I say, quantitative differences in intrinsic value. Surely, to destroy the earth as a whole would be more dreadful than swatting a fly, or even eradicating flies as a species. Thus, scientists even hesitate to destroy the last smallpox virus which they have isolated in a laboratory somewhere.

We must remember that in the mystical state we identify not only with each manifestation of being, but with *being as such*—that is with being as a whole in all of its interactive manifestions. Our concern is thus not just with this or that manifestation of being (but of course that too), but with each manifestation in its particular niche within the interactive whole of reality. Individual moral decisions can never be taken apart from a compassionate concern for the whole. That does not lead to a particular moral judgment in one or another specific situation. Rather, it leads to a different way of viewing a particular situation and thus making moral decisions about it—a view *sub specie aeternitatis* (from the eternal point of view), a wider and more compassionate state of being and character that has emerged from the various stages in the spiritual journey of life. It is a transformed and transparent spiritual state of love and compassion that weighs such decisions in the light of a deep concern for all being.

Let us return to our brief outline of deep ecology's ethical perspective. Its proponents are arguing for an ethics that is spiritually based rather than grounded on an intellectual principle.

> When perception is sufficiently changed, respectful types of conduct seem "natural," and one does not have to belabor them in the language of rights and duties. Here, finally, we reach the point of "paradigm change." What brings it about is not exhortation, threat, or logic, but a rebirth of the sense of wonder that in ancient times gave rise to philosophers but is now more often found among field naturalists.[11]

Such an experience, Naess claims, is based on an "identification" with human and nonhuman modes of reality and constitutes a kind of self-realization of a wider or bigger self. "We ourselves, as human beings, are capable of identifying with the whole of existence."[12]

> How do we develop a wider self?...The self is as comprehensive as the totality of our identifications. Or, more succinctly: Our Self is that with which we identify. The question then reads: How do we widen identification?[13]
>
> Self-realization cannot develop far without sharing joys and sorrows with others, or more fundamentally, without the development of the narrow ego of the small child into the comprehensive structure of a Self that comprises all human beings. The [deep] ecological movement—as many earlier philosophical movements—takes a step further and asks for a development such that there is a deep identification of individuals with all life.[14]

In other words, this spiritually based ethics occurs when a person or group integrates themselves into the larger story of the universe or, if you prefer, when they integrate the story of the universe into their own story, thereby transforming themselves and their behavior toward encompassing reality.

Rather than an appeal to theoretical principles, the transformational ethical model which flows from the new cosmology is an ethics based on actual experience and persistent character formed in the light of reality itself. Like the Sermon on the Mount, it is an ethics of love and compassion rather than rules and duties. Such an ethics involves real caring relationships with others rather than individual judgments about moral acts. Physicists Menas Kafatos and Robert Nadeau draw precisely this moral conclusion from their exploration of contemporary physical theory.

> Sacrifice on this order requires a profound sense of identification with the "other" that operates at the deepest levels of our emotional lives. And it is this sense of identification which has, quite obviously, always been one of the primary challenges and goals of religious thought and practice…. Central to this vision would be a cosmos rippling with tension evolving out of itself endless examples of the awe and wonder of its seamlessly interconnected life. And central to the cultivation and practice of the spiritual pattern of the community would be a profound acceptance of the astonishing fact of our being.[15]

Ecological thinker Warwick Fox specifies three senses of identification and thus self-realization—the personal, ontological, and cosmological. What deep ecology refers to as "identification" I have called the mystical experience of and identification with "the miracle of being." My approach in this book approximates a combination of what Fox calls the ontological and cosmological senses of identification.

> Ontologically based identification refers to experiences of commonality with all that is that are brought about through deep-seated realization of the fact that

things are.... that things are impresses itself upon some people in such a profound way that all that exists seems to stand out as foreground from a background of nonexistence, voidness, or emptiness—a background from which this foreground arises moment by moment.

Cosmologically based identification refers to experiences of commonality with all that is that are brought about through deep-seated realization of the fact that we and all other entities are aspects of a single unfolding reality. This realization can be brought about through the empathic incorporation of any cosmology (i.e., any fairly comprehensive account of how the world is) that sees the world as a single unfolding process—as a "unity in process," to use Theodore Roszak's splendid phrase.[16]

In summary, then, an ethics that flows from the new cosmology is first of all theocentric, not in the sense of divine orders and commands nor in the sense of a moral casuistry that resolves every ethical dilemma before it arises. Rather, it is theocentric in the sense of a mystical identification in wonder with being in all its diverse modes and manifestations. The mystical apprehension of and identification with the mysterious and surprising coming-into-being of new forms of nature includes me and my consciousness. Thus, I am freed in such an apprehension to become myself in a process of self-actualization.

Second, insofar as we encounter nature in its particularity and as a whole as a mode and manifestation of the ultimate power-to-be, transformational ethics is biocentric. Nature is encountered as intrinsically valuable and worthy of ethical concern.

Third, from the perspective of transformational ethics, the ultimate ground of ethics is the fundamental worldview (hermeneutic) held by an individual or group. And since such interpretive understandings found and pervade cultures, there are social and cultural structures which often condition and shape our moral behavior. For example, because natural resources are utilized for production and consumption in the political economy of modern industrial societies, then we are structurally conditioned to ethically use (and all too often overuse) them for our anthropocentric purposes. Insofar as we live in such a society, we all are forced by society itself to participate in such damaging behavior even though it may not be termed "ethical" or "unethical" in any explicit sense. This is important for transformational ethics because it means that we are often blinded to such structural implications by the wider attitude embedded in our culture. Rather than conceiving of morality as merely individual judgment and behavior, we need to recognize that it becomes embedded in cultural processes and institutions. Furthermore, it means that serious transformational ethics must ultimately come to terms with these structural moral concerns in our political economy.

Fourth, this ethics is not based simply or solely on individual judgments, but always finds its context in the identification and caring for and with other existing entities, including but not limited to other human beings. Ethical decisions here flow from feelings of care and trust rather than intellectual principles; and such decisions are made socially with others rather than exclusively by individuals.

Fifth, this ethical perspective entails moving away from focussing on particular and isolated individual moral acts to moral activity embedded in real temporal and cultural contexts.

Sixth, this is not ethics in the sense of an appeal to rationally demonstrated abstract principle. Such an attempt to philosophically "justify" particular moral judgments implicitly assumes the fundamental importance and ultimacy of such intellectual principles, and those principles are in fact never founded and self-justified! Whatever principle is placed in this privileged position can always be morally questioned. In other words, such a procedure reveals a faith in rational demonstration as a means of grounding morality instead of God or the wider ground of being itself.

Seventh, this is a form of virtue ethics in that character is essential to moral behavior—that is, behavior shaped by what we ultimately care about. Moral decisions are not only shared with other caring people with whom one is personally connected in mutual trust, but are also based on feelings of love and compassion, on character and virtue that develops from a deep and positive affection and concern for others as holy manifestations of the mysterious power-to-be. As the famous Jewish thinker Martin Buber said, others are not "its," but "thou's" or "images of God." Christian theologian and ethicist James Gustafson contends that feeling is essential to his theocentric morality in that it moves us to do it.

> Piety, evoked by the senses of dependence, gratitude, obligation, remorse and repentance, possibility, and direction, is thus an essential aspect of theocentric morality. Piety involves feelings or affectivity. Theocentric piety and fidelity shape our sense of the value of the human in relation to the patterns and processes of interdependence in which we live. It "moves" (motivates) our participation as well as grounds some of our reasons for it.[17]

Last, this ethics does not claim to "solve" all ethical issues. Resolving all such issues by reducing them to a fundamental and universal moral principle is a project that banishes ambiguity and the need to take personal responsibility in favor of an intellectual and abstract certainty. Why else would we do such a thing? This ethics, to the contrary, acknowledges that there is no such certainty, that the future is open and that our moral decisions are in a certain sense contributions to what those future states will come to be. By reducing all ethical quandaries to an intellectual principle or theoretical instance, we neatly avoid having to decide, having to risk such a commitment. As Kierkegaard might have put it, we transform a subjective truth into an objective one, thereby escaping the fact that morality is always "an objective uncertainty." Instead of rushing in to solve all moral questions by stuffing them into a theoretical box that frees us from the need to take responsibility, the ambiguity of our lives and the openness of the future are left intact. Gustafson puts this well:

> To see ourselves from a theocentric perspective is to see and feel ourselves in a condition of ambiguity. This condition cannot be totally relieved, from such a perspective, by the construction of an ideal ethical theory that provides almost absolute certainty about ourselves and our actions.[18]

Rather than rational principles, then, this ethical model entails transformed and transparent, self-actualizing and caring persons who do their best together to make moral decisions in the light of their identification with and compassionate concern for holy being in all of its myriad forms.

Culture

Descartes uncoupled nature from god, thereby divorcing it from spiritual epiphany and moral compunction. Science and religion went their separate ways. In the meantime, that stripped-down nature which was thought to be merely a utilitarian resource with no intrinsic value was turned over to the tender mercies of a manipulative technology and omnivorous market. The decoupling of nature and spirit led in the modern industrial societies that followed to a ravaged nature (now thought of as a "natural resource" with no intrinsic value) and an increasingly demoralized human culture cut off from any spiritual depth or orientation.

Perhaps the most all-embracing and pervasive result of the intellectual revolution I have traced in this book may be a complete shift in worldview. The core of my argument here has been that a change in interpretive understanding entails a spiritual transformation in both the individuals and—when effective—in the cultures in which it occurs. And because such hermeneutical transformations "found" and pervade human worlds or cultures, the intellectual revolution we are witnessing may result in a cultural shift in the symbolic framework of our culture labelled variously "postmodern" or "ecozoic." Just one piece of this change might be a shift from "seeing" nature "as" a mere resource to "seeing" it "as" the body of God or the location and occasion for epiphany. Such a shift would surely not be inconsequential.

Another aspect of such a cultural transformation may be a change in the intellectual landscape in which we understand science and religion. There seems to be a widespread yearning within our culture to heal the breach between science and religion, nature and spirit, and reason and faith because these represent fundamental and unavoidable needs within ourselves. As theologian Margaret Wertheim has said,

> many people deeply desire a resolution between science and spirituality. Many are tired of being told that they must choose between faith and reason. They want both forces in their lives.... The psychological need to bridge the divide between science and spirituality is great in our age.[19]

A healing of this deep separation of science and religion, the head and the heart, may now be possible. Postmodern understandings of religion and science together with the unfolding of the new scientific cosmology in the past forty years have provided a new lens through which to see nature and our place within it. Religious understanding is not explanatory science but an interpretive understanding of the meaning of being most basically and directly made available in creation mythologies. The ultimate and transcendent reality is made manifest through those stories to the mystical experience of wonder, and linked to each and every aspect of creation. Thus, nature is a manifestation of holy being, the home of the spirit. Physicist Wolfgang Pauli put it this way.

> Contrary to the strict division of the activity of the human spirit into separate departments—a division prevailing since the nineteenth century—I consider the ambition of overcoming opposites, including also a synthesis embracing both rational understanding

and the mystical experience of unity, to be the mythos, spoken and unspoken, of our present day and age.[20]

The new scientific cosmology offers both a scientifically accurate understanding of the evolution of the entire universe and a spiritual vision of a wider order of being to which we belong and from and in which we find our genesis and home. God is neither nature itself nor located apart from it, but is available to mystical experience within it. Paraphrasing preservationist John Muir, everything seems connected to everything else. God or ultimate reality is not nature, but the "miracle of being" which each and every aspect of nature displays and which we encounter in wonder and awe.

Old and tired worldviews and cultural attitudes toward life are not overcome by dispute and argument any more than is a person's fundamental faith in life. Rather than being disproved, they simply dissolve when they no longer meet significant human needs and longings, and thus make room for new perspectives. The unholy breach between nature and spirit, science and religion, head and heart may be in process of dissolving. If so, then we may have a chance to live whole and balanced lives again, lives in which we can use our heads to gain a bit of security in life as well as open our hearts to the deep and holy reality unfolding within and without us. That may be one of the most important promises of the Promised Land toward which—sometimes in pain and confusion and sometimes in hope and high expectation—we have been making our way.

Notes

Preface

    1. Virginia Stuart, "Physics Can Be Simply Divine," *Nashua Telegraph* (Science and Technology Section), April 29, 1996, 10.

Introduction

    1. Richard Leakey, *Origins Reconsidered* (New York: Anchor, 1992), 339-40.

    2. James Irwin, *The Home Planet*, Kevin Kelley, ed. (Reading MA.: Addison Wesley Co, 1988), 38.

    3. Olag Makarov, *The Home Planet*, Preface.

    4. Edgar Mitchell, *The Home Planet*, Endpiece and 52.

    5. Kevin W. Kelley, *The Home Planet*, Introduction.

    6. John Fowles, "Seeing Nature Whole," in Susan J. Armstrong and Richard G. Botzler, eds., *Environmental Ethics: Divergence and Convergence* (New York: McGraw Hill, 1993), 141.

    7. Ralph Waldo Emerson, "Nature," in *The Portable Emerson*, Carl Bode, ed. (New York: Penguin Books, 1981), 10.

    8. William Blake, "The Marriage of Heaven and Hell," in June Singer, *The Unholy Bible* (New York: G. P. Putnam's Sons, 1970), plate 14.

    9. Vaclav Havel, *Letters to Olga*, Paul Wilson, trans. (New York: Knopf, 1988), 365-66.

    10. Vaclav Havel, "Creating a New Vision," *EarthLight*, Spring, 1995, 6.

    11. Richard Eckersley, "The West's Deepening Cultural Crisis," *The Futurist*, Nov./Dec., 1993, 10.

    12. Burton Mack, *Who Wrote the New Testament?* (San Francisco: HarperSanFrancisco, 1995), chapter 1.

13. David Bollier, "Who 'Owns' the Life of the Spirit?", *Tikkun*, Jan/Feb., 1994, 89.

14. Vaclav Havel, "The Post-Communist Nightmare," *New York Review of Books*, May 27, 1993, 10.

15. Vaclav Havel, "Civilization's Thin Veneer," *Harvard Magazine*, July/August 1995, 34, 66.

16. Herman E. Daly and John B. Cobb, Jr., *For the Common Good: Redirecting the Economy Toward Community, the Environment, and a Sustainable Future* (Boston: Beacon Press, 1989), 373-75.

17. "Cosmology" (from the Greek) means a theory of the whole or understanding of the universe in its entirety. It has often been used to refer simply to the origin of the universe; but since the universe is in fact an emerging process, I shall use the term to mean an understanding (in narrative form) of the whole, evolving reality we call the universe.

18. Paul Davies, *God and the New Physics* (New York: Touchstone, Simon & Schuster, Inc., 1983), 229.

19. Joseph Campbell, "The Mythological Dimension," *Historical Atlas of World Mythology*, vol. 1 (New York: Perennial Library, 1988), 8.

Chapter One

1. Brian Swimme and Thomas Berry, *The Universe Story* (San Francisco: HarperSanFrancisco, 1992), 242.

2. See for example Lynn White's classic essay, "The Historical Roots of Our Ecological Crisis," *Science*, vol. 155, no. 3767, March 10, 1967.

3. John Hick, *An Interpretation of Life* (New Haven and London: Yale University Press, 1989), 301.

4. Paul Brockelman, *The Inside Story: A Narrative Approach to Religious Understanding* (Albany, NY: SUNY Press, 1992).

5. Langdon Gilkey, from the transcript of his testimony at the trial of 1981, in *Religion and the Natural Sciences: the Range of Engagement*, James E. Huchingson, ed. (New York: Harcourt Brace Jovanovich College Publishers, 1993), 62, 64.

6. Ian G. Barbour, "Creation and Cosmology," in *Cosmos as Creation*, Ted Peters, ed. (Nashville: Abingdon Press, 1989), 146.

7. See Paul Tillich, *The Dynamics of Faith* (New York: Harper & Row, 1958), ch. 3.

8. David Klemm, *Hermeneutical Inquiry*, vol. 1 (Atlanta, GA: Scholars Press, 1986), 42.

9. From a letter written by Joseph Epes Brown, quoted in Huston Smith, *The World's Religions* (San Francisco: Harper, 1991), 379.

10. T. C. McCluhan, *The Way of the Earth: Encounters with Nature in Ancient and Contemporary Thought* (New York: Simon & Schuster, 1994), 41.

11. Leonardo Boff, *Ecology and Liberation: A New Paradigm* (Maryknoll, NY: Orbis Books, 1995), 143.

12. G. Filoramo, *A History of Gnosticism* (Oxford: Blackwell, 1990), 44, 53.

13. Quoted in Filoramo, 39.

14. "The Huai-Naz Tzu: the Creation of the Universe," from Barbara C. Sproul, *Primal Myths: Creation Myths from Around the World* (San Francisco: HarperSanFrancisco, 1979), 206-207.

15. Burton L. Mack, *Who Wrote the New Testament?* (San Francisco: HarperSanFrancisco, 1995), 30.

16. Leonard J. Biallas, *Myths: Gods, Heroes, and Saviors* (Mystic CT: Twenty-Third

Publications, 1989), 39.

17. Max Black, *Models and Metaphors: Studies in Language and Philosophy* (Ithaca, NY: Cornell University Press, 1962), 41.

18. Nikos Kazantzakis, *Report to Greco*, P. A. Bein, trans. (New York: Simon & Schuster, 1965), 291-92.

19. Clifford Geertz, "Religion as a Cultural System," *The Interpretation of Cultures* (New York: Basic Books, 1973), 89.

20. Mircea Eliade, *The Scared and the Profane*, Willard Trask, trans. (New York: Harcourt Brace Jovanovich, 1959), 95.

21. As Joseph Campbell puts it, "The first function of a mythology is to waken and maintain in the individual a sense of wonder and participation in the mystery of this finally inscrutable universe." Joseph Campbell, "The Mythological Dimension," *Historical Atlas of World Mythology*, vol. 1 (New York: Perennial Library, 1988) 8.

Chapter Two

1. Holmes Rolston, III, *Environmental Ethics* (Philadelphia: Temple University Press, 1988), 230.

2. Eight to twelve billion years if the recent (October, 1994) Hubble Space Telescope measurements are confirmed, which looks increasingly unlikely.

3. Rupert Sheldrake, *The Rebirth of Nature: The Greening of Science and God* (New York: Bantam Books, 1992), 182.

4. Alan Lightman, *Ancient Light: Our Changing View of the Universe* (Cambridge: Harvard University Press, 1991), 100.

5. Timothy Ferris, *Coming of Age in the Milky Way* (New York: Anchor Books, 1988), 286-89 and 384.

6. Menas Kafatos and Robert Nadeau, *The Conscious Universe: Part and Whole in Modern Physical Theory* (NY: Springer Verlag, 1990), 175.

7. J. Sturrock, *Structuralism and Since* (Oxford: Oxford University Press, 1979), 164.

8. Timothy Ferris, *The Whole Shebang: A State of the Universe(s) Report* (New York: Simon and Schuster, 1997), 17, 196, 311.

9. Timothy Ferris, *Coming of Age in the Milky Way*, 384-387.

10. John Haught, "Religious and Cosmic Homelessness: Some Environmental Implications," in *Liberating Life: Contemporary Approaches to Ecological Theology*, Charles Birch, Willian Eakin, and Jay McDaniel, eds. (Maryknoll: N.Y.: Orbis Books, 1990), 173.

11. Ilya Prigogine and Isabelle Stengers, *Order Out of Chaos: Man's New Dialogue with Nature* (New York: Bantam Books, 1984), 92.

12. John Gribben, *In the Beginning: After Cobe and Before the Big Bang* (Boston: Little, Brown and Co., 1993), 246-7.

13. As we shall see in chapter 5, this creative emergence also characterizes each person's experience to himself. That is, our actions in the present transcend or aim toward futures which they bring into a new present. Thus, each moment in our consciousness picks up where earlier moments left off but at the same time goes beyond them. In a real sense, then, God or the mysterious power-to-be which so characterizes the universe as a whole also shows up in the consciousness of each of us and the cultural history of all of us.

14. Brian Swimme and Thomas Berry, *The Universe Story* (San Francisco: HarperSanFrancisco, 1992), 74.

15. Kafatos and Nadeau, *The Conscious Universe*, 178.

16. Thomas Berry, "The Gaia Theory: Its Religious Implications," unpublished paper, 18-19.

17. Sallie McFague, *The Body of God* (Minneapolis: Fortress Press, 1993), 105.

Chapter Three

1. For example, see Thomas V. Morris, ed., *The Concept of God* (Oxford: Oxford University Press, 1987), 4-5.

2. Paul Ricoeur, "Philosophy of Will and Action," in *Phenomenology of Will and Action,* Straus and Griffith, eds., (Pittsburgh: Duquesne University Press, 1967), 17.

3. Wayne Proudfoot, *Religious Experience* (Los Angeles: University of California Press, 1985), 196.

4. Maurice Merleau-Ponty, *Phenomenology of Perception,* (New York: Humanities Press, 1962), vii.

5. William Pollard, *The Mystery of Matter,* (U.S. Atomic Energy Commission, Office of Information Services, 1974), 54.

6. Leonardo Boff, *Ecology and Religion: A New Paradigm* (Maryknoll, NY: Orbis Books, 1995), 138.

7. Rudolph Otto, *The Idea of the Holy,* John W. Harvey, trans. (New York: Oxford University Press, 1958).

8. Vaclav Havel, *Temptation,* Marie Winn, trans. (New York: Grove Press, 1989), 29.

9. John F. Haught, "Religious and Cosmic Homelessness," in *Liberating Life: Contemporary Approaches to Ecological Theology,* Birch, Eakin, McDaniel, eds. (Maryknoll, N.Y.: Orbis Books, 1990), 172, 175, 178.

10. Dennis Overbye, *Lonely Hearts of the Cosmos* (New York: Harper Collins, 1991), 3.

11. Loren Eisley, *The Firmament of Time* (New York: Atheneum, 1960), 171.

12. Norman Malcolm, *Ludwig Wittgenstein: A Memoir* (London: 1958) 70.

13. Ludwig Wittgenstein, *Tractatus Logico-philosophus* (London: Routledg and Kegan Paul, 1974), 6.44.

14. Marcus Borg, *Meeting Jesus for the First Time* (San Francisco: HarperSanFrancisco, 1994), 61.

15. Kafatos and Nadeau, *The Conscious Universe,* 180.

16. John Wheeler in an interview with Timothy Ferris, quoted by Ferris, *Coming of Age in the Milky Way* (New York: Anchor, 1988), 387.

17. Barbara C. Sproul, *Primal Myths: Creation Myths Around the World* (San Francisco: HarperSanFrancisco, 1991), 21.

18. Quoted by Robert Moss, in "We Are All Related," Parade Magazine, Oct. 11, 1992, 8.

19. Sallie McFague, *The Body of God* (Minneapolis: Fortress Press, 1993), 194.

20. Mircea Eliade, *The Sacred and the Profane,* W. Trask, trans. (New York: Harper & Row, 1961), 12.

21. Meister Eckhart, *Breakthrough: Meister Eckhart's Creation Spirituality* (New York: Doubleday, 1980), Sermon 4, 84.

22. Thomas Berry, *Befriending the Earth: A Theology of Reconciliation Between Humans and the Earth* (Mystic, CN: Twenty-Third Publications, 1991), 11.

23. Albert Schweitzer, *Out of My Life and Thought,* C. I. Campion, trans. (New York: Henry Holt & Co., 1933), 186.

Chapter Four

1. Walter Brueggemann, "Transforming the Imagination," *Books and Religion,* Spring,

1992, 10.

2. Martin Heidegger, *Being and Time*, John Macquarrie and Edward Robinson, trans. (New York: Harper & Row, 1962), 26.

3. See Thomas Aquinas, *On Being and Essence*, A. Maurer, trans. (Toronto: Pontifical Institute of Medieval Studies, 1949), 50; and *Summa Contra Gentiles*, Book 1, Chap. 26 ("That God is Not the Formal Being of All Things").

4. Martin Hedegger, *Being and Time*, 29.

5. Gordon D. Kaufman, "Mystery, Theology, and Conversation: Convocation Address, 1991," in *Harvard Divinity Bulletin*, 1991-1992, vol. 21, No. 2, 12. Also, see his recent *In Face of Mystery* (Cambridge MA: Harvard University Press, 1993), 60-61.

6. See *Face of Mystery*, 56, 328, 330, 331.

7. Quoted in Andrew Harvey and Anne Baring, eds, *The Mystic Vision* (San Francisco: HarperSanFrancisco, 1995), 19.

8. F. L. Cross and E. A. Livingstone, eds., *The Oxford Dictionary of the Christian Church* (Oxford: Oxford University Press, 1997), 1213.

9. Dionysius the Areopagite, *Letter to Dorothy the Deacon*, quoted in Andrew Harvey and Anne Baring, eds. (San Francisco: HarperSanFrancisco, 1995), 91.

10. Since creation did not originate in time and thus was eternal in this Hellenistic-Roman context, God or the first principle was thought to be a sort of rational or dialectical ground of the universe for the Neo-Platonists and a Final Unmoved Cause for the Aristotelians. St. Thomas Aquinas later added to these two notions of God's creating activity the apparently biblical conception of God as a first cause creator of a universe that had a beginning.

11. I. Newton, *Principia Mathematica* 1686, I (The Motion of Bodies); II (The System of the World), Cajori, ed. (Berkeley: University of California, 1966), 545.

12. *Luther's Works*, vol. 37 J. Pelikan, ed. (St. Louis: Concordia Publishing House, 1955 ; and Philadelphia: Fortress Press, 1955), 58.

13. Robert McCrum, "My Old and New Lives: What happens when, in the course of a night, your life is changed forever?", *The New Yorker*, May 27, 1996, 117.

14. Robert Jastrow, *God and the Astronomers* (New York: W. W. Norton & Co., Inc., 1992), 123-4.

15. Sallie McFague, "Imaging a Theology of Nature," in *Liberating Life: Contemporary Approaches to Ecological Theology*, Birch, Eakin, McDaniel, eds. (Maryknoll, NY: Orbis Books, 1990), 218.

16. On panentheism see footnote no. 8. above.

17. Arthur Peacocke, *Theology for a Scientific Age* (Minneapolis: Fortress Press, 1993), 105.

18. Paul Davies, *The Mind of God* (New York: Touchstone Books, 1992).

19. Stephen Hawking, *A Brief History of Time: From Big Bang to Black Holes* (New York: Bantam Books, 1988), 136.

20. Hawking, *A Brief History of Time*, 174.

21. Renee Weber, *Dialogues with Scientists and Sages* (London: Arkana, 1990), 6.

22. Fyodor Dostoevsky, *A Raw Youth*, Constance Garnett, trans. (London: William Heineman Ltd., 1950), 351.

23. R. H. Blyth, *Haiku*, vol. 1 (San Francisco: Heian International, Inc.), 31.

Chapter Five

1. For an explanation of this form of understanding, see pp. 22ff in chapter 1.

2. See Saint Bonaventura, *The Mind's Road to God*, George Boaz, trans. (Indianapolis: Library of Liberal Arts, 1953).

3. For a brief description of this philosophical approach, see pages 66-7 in chapter 3.

4. Martin Heidegger, *Poetry, Language, Thought*, Albert Hofstadter, trans. (New York: Harper and Row, 1971), pp. 116-17.

5. Martin Heidegger, *Being and Time*, John Macquarrie and Edward Robinson, trans. (New York: Harper and Row, 1962), 33, and 67-70.

6. Vaclav Havel, *Letters to Olga*, Paul Wilson, trans. (New York: Knopf, 1988), 365-6.

7. Iris Murdoch, "The Sublime and the Good," *Chicago Review*, 13 (Fall 1959): 51.

8. David Kolb, *The Critique of Pure Modernity* (Chicago: University of Chicago Press, 1986), 175.

9. Erazim Kohak, *The Embers and the Stars: A Philosophical Enquiry into the Moral Sense of Nature* (Chicago: University of Chicago Press, 1984), 60-61.

10. Sallie McFague, *The Body of God*, 123.

11. Marty Kaplan, *Time*, June 24, 1996, 62.

12. Karsten Harries, *The Ethical Function of Architecture* (Boston: MIT Press, 1997), 133.

13. Holmes Rolston III, *Environmental Ethics: Duties to and Values in the Natural World* (Philadelphia: Temple University Press, 1988), 235.

Chapter Six

1. Initially, this was accomplished by allegorizing the Greek myths and gods as well as the biblical stories as Christianity spread throughout the Roman Empire. Those myths were interpretively glossed to be actually philosophical systems and perspectives, especially Stoicism with its Logos doctrine and later Neo-Platonism in which earthly reality was seen as a secondary reflective image of the interrelated primary reality of ideas in the mind of God. For example, the early Christian father, Origen, interpreted biblical stories and events on three allegorical levels, thereby transforming them into philosophical understanding acceptable to the upper classes of the Hellenistic and Roman culture.

2. Usually referred to as "the kingdom of God." See also the Sermon on the Mount, the Pentecostal events after the death of Jesus, and the Pauline notion of living in the spirit.

3. Rachel Carson, *The Sense of Wonder*, Charles Pratt, ed. (New York: Harper & Row, 1965), 88.

4. See for example, John Hick, *An Interpretation of Religion* (New Haven: Yale University Press, 1989), ch. 3.

5. Lao Tsu, *Tao Te Ching*, Gia-Fu Feng and Jane English, trans. (New York: Vintage, 1972), ch. 28.

6. Danu Baxter, four-and-a-half years old, quoted in Sallie McFague, *The Body of God: An Ecological Theology* (Minneapolis, MN: Augsburg Fortress, 1993), 130.

7. Steven Rockefeller, "The Wisdom of Reverence for Life," in *The Greening of Faith: God, the Environment, and the Good Life*, J. Carroll, P. Brockelman, and M. Westfall, eds. (Hanover, NH: University Press of New England, 1996), 46.

8. Augustine, *The Confessions of Saint Augustine*, Edward B Pusey, trans. (New York: The Modern Library, 1949), 3.

9. Dale Cannon, *Six Ways of Being Religious* (Belmont CA.: Wadsworth Publishing Co., 1996), 60. See also, 43.

10. Frederick Streng, *Understanding Religious Life* (Encino CA: Dickenson Pub. Co., 1976), 66-67.

11. See Joseph Epes Brown, *The Sacred Pipe* (Baltimore: Penguin Books Inc., 1971), chap. 3.

12. Lewis R. Rambo, *Understanding Religious Conversion* (New Haven: Yale University Press, 1993), 129.

13. Roger S. Gottlieb, ed., *This Sacred Earth: Religion, Nature, and Environment* (New York: Routledge, 1996), 8.

14. Denise Lardner Carmody and John Tully Carmody, *Mysticism: Holiness East and West* (New York: Oxford University Press, 1996), 298.

15. Carmody and Carmody, *Mysticism*, 13.

16. William James, *Varieties of Religious Experience* (1902) (London and New York: Collins and Mentor Books, 1960), 268-70; quoted in John Hick, *An Interpretation of Religion* (New Haven and London: Yale University Press, 1989), 302.

17. Arne Naess quoted in Stephan Bodian, "Simple in Means, Rich in Ends: An Interview with Arne Naess," in *Deep Ecology for the Twenty-first Century*, George Sessions, ed. (Boston: Shambhala Publications, Inc., 1995), 30.

18. Augustine, Sermon 227, *Fathers of the Church*, 196; quoted in Margaret Miles, *Carnal Knowing: Female Nakedness and Religious Meaning in the West* (Tunbridge Wells, Kent, England: Burns & Oates, 1992), 41 (italics mine).

19. Ralph Wendell Burhoe, "Attributes of God in an Evolutionary Universe," in *Religion and the Natural Sciences: The Range of Engagement*, James E. Huchingson, ed. (New York: Harcourt, Brace, Jovanovich College Press, 1993), 305.

20. Soren Kierkegaard, *Fear and Trembling and the Sickness Unto Death*, Walter Lowrie trans. (Princeton: Princeton University Press, 1974), 213.

21. David Burrell, *Exercises in Religious Understanding* (Notre Dame: University of Notre Dame Press, 1974), 220.

22. James McClendon, *Biography as Theology* (Philadelphia: Trinity International Press, 1990), 20. Also, see Stanley Hauerwas, *Character and the Christian Life: A Study in Theological Ethics* (San Antonio, TX.: Trinity University Press, 1979).

23. Thomas Moore, *The Care of the Soul* (New York: Harper Collins, 1992), 253.

24. Gordon Kaufman, *In Face of Mystery* (Cambridge: Harvard University Press, 1993), 58.

25. See Soren Kierkegaard, *Either/Or* in *A Kierkegaard Anthology*, R. Bretall, ed. (New York: Random House, 1946), 97ff.

26. Arthur Peacocke, *Theology for a Scientific Age* (Minneapolis: Fortress Press, 1993), 325.

27. Steven Rockefeller, "Reverence for Life," 52.

28. Quoted in Hick, *An Interpretation of Religion*, 316.

29. Alice Walker, *The Color Purple* (New York: Pocket Books, 1983), 247.

30. Ken Wilbur, *Sex, Ecology, Spirituality: The Spirit of Evolution* (Boston: Shambala, 1995), 291.

31. Warwick Fox, *Toward a Transpersonal Ecology: Developing New Foundations for Environmentalism* (Albany, NY: SUNY Press, 1995), 257-8.

Chapter Seven

1. Thomas Berry, *The Dream of the Earth* (San Francisco: Sierra Club Books, 1988), 124.

2. For more on this, see Paul Brockelman, *The Inside Story: A Narrative Approach to Religious Understanding and Truth* (Albany: SUNY Press, 1992), chs. 3 and 4.

3. Paula Butturini, "For Monks, A Communion of Heart: Benedictines, Buddhists

Form Spiritual Bond," *The Boston Sunday Globe*, June 9, 1996, 12.

4. Warwick Fox, *Toward a Transpersonal Ecology: Developing New Foundations for Environmentalism* (Albany, NY: SUNY Press, 1995), 254.

5. Timothy Ferris, *The Whole Shebang: A State-of-the-Universe(s) Report* (New York: Simon and Schuster, 1997), 302.

Chapter Eight

1. Menas Kafatos and Robert Nadeau, *The Conscious Universe: Part and Whole in Modern Physical Theory* (New York: Springer Verlag, 1990), 186.

2. Thomas Berry, "The Viable Human," in *Deep Ecology for the Twenty-First Century*, George Sessions, ed. (Boston: Schambhala Publications, Inc., 1995), 12,18.

3. Angela Tilby, *Soul: God, Self and the New Cosmology* (New York: Doubleday, 1993), 109.

4. Sallie McFague, *The Body of God: An Ecological Theology* (Minneapolis: Fortress Press, 1993), 111.

5. Allen W. Wood, *Hegel's Ethical Thought* (Cambridge: Cambridge University Press, 1993), 21.

6. Ken Westphal, "The Basic Context and Structure of Hegel's Philosophy of Right," in *The Cambridge Campanion to Hegel*, F. C. Beiser ed. (Cambridge: Cambridge University Press, 1993), ch. 8.

7. Albert Schweitzer, *Out of My Life and Thought*, C. T. Campion, trans. (New York: Henry Holt & Co., 1933), 188.

8. Steven Rockefeller, "Reverence for Life," in *The Greening of Faith: God, the Environment, and the Good Life*, J. Carroll, P. Brockelman, and M. Westfall, eds. (Hanover, NH: University Press of New England, 1996), 57.

9. Arne Naess, "The Deep Ecological Movement: Some Philosophical Aspects," in *Deep Ecology for the Twenty-first Century*, 68.

10. Arne Naess, from a speech delivered in Australia in 1984, quoted by Warwick Fox, *Toward a Transpersonal Ecology: Developing New Foundations for Environmentalism* (Albany, NY: SUNY Press, 1995), 219.

11. John Rodman, "Four Forms of Ecological Consciousness Reconsidered," in *Deep Ecology*, 127.

12. Arne Naess, quoted in Stephen Bodian, "Simple in Means, Rich in Ends: An Interview with Arne Naess," in *Deep Ecology*, 27.

13. Arne Naess, "Identification as a Source of Deep Ecological Attitudes," in *Deep Ecology*, Michael Tobias, ed. (San Diego: Avant Books, 1985), 261. Quoted in Fox, 230.

14. Naess, "Notes on the Methodology of Normative Systems," *Methodology and Science* 10 (1977), 71. Quoted in Fox, 230.

15. Kafatos and Nadeau, *The Conscious Universe*, 186,187,188.

16. Warwick Fox, *Transpersonal Ecology*, 250, 251, 252.

17. James Gustafson, *Ethics from a Theocentric Perspective*, 2 volumes (Chicago: University of Chicago Press, 1984), vol. 2, 185.

18. Gustafson, Ethics, 289-90.

19. Margaret Wertheim, "Physics, Faith, and Feminism: 1995-96 J. K. Russell Fellowship Lecture," *CTNS Bulletin*, vol. 16, no. 2, Spring, 1996, 7.

20. Quoted by Kafatos and Nadeau, *The Conscious Universe*, 188.

Index